# Sustainable Living with Environmental Risks

Nobuhiro Kaneko • Shinji Yoshiura
Masanori Kobayashi
Editors

# Sustainable Living with Environmental Risks

*Editors*
Nobuhiro Kaneko
Yokohama National University
Yokohama
Japan

Shinji Yoshiura
Yokohama National University
Yokohama
Japan

Masanori Kobayashi
Yokohama National University
Yokohama
Japan

ISBN 978-4-431-54803-4 ISBN 978-4-431-54804-1 (eBook)
DOI 10.1007/978-4-431-54804-1
Springer Tokyo Heidelberg New York Dordrecht London

Library of Congress Control Number: 2014930336



Printed on acid-free paper

Springer is part of Springer Science+Business Media (www.springer.com)

# Foreword

Managing risks and promoting sustainability are imperative for the international community to build a sustainable society on our planet. To prompt changes in our behaviours and business models and to facilitate the socio-economic transformation necessary for a sustainable world, we must empower people who can take the lead in supporting the transition to a system that is more environmentally sound and enriching in terms of human well-being.

Yokohama National University (YNU) strives to promote international partnerships as a means of advancing education and research. The Leadership Programme in Sustainable Living with Environmental Risk (the SLER programme) is one such initiative that YNU has promoted over the years. Since 2009, the SLER programme has been supported by the Japanese Ministry of Education, Culture, Sports, Science and Technology (MEXT), and the Japanese Science and Technology Promotion Agency (JST). It is spearheaded by the YNU Graduate School of Environment and Information Sciences, and is considered one of the key initiatives at our university. Prof. Shin Morishita, Dean of the YNU Graduate School of Environment and Information Sciences, has championed the SLER programme as Programme Leader, and Prof. Nobuhiro Kaneko has steered SLER education and research activities as Programme Coordinator together with other faculty and supporting staff.

This textbook, *Sustainable Living with Environmental Risks*, is a manifestation of the progress and achievements made by the SLER programme. I would like to express my gratitude and appreciation to MEXT, JST, and all those in YNU, our partner universities, and elsewhere, who have collaborated to support the programme and the publication of this textbook. I trust that this textbook will provide a useful knowledge base and tool for future students and practitioners to comprehend the mounting challenges we all face and to explore innovative approaches to manage risk effectively and promote sustainability. We expect graduates of the SLER programme at YNU and partner universities to bring both their expertise and their compassion to bear in leading the transformation that will bring about a sustainable society and world.

Global sustainability cannot be achieved by a single stakeholder group or country. We must work more closely together and enhance the impact of our educational and research activities to develop the knowledge and skills essential for future environmental leaders. We must forge ahead with our endeavours and capitalize upon the SLER initiative's achievements. In this way, YNU will continue to play a leading role in developing environmental leaders for a sustainable world.

Kunio Suzuki
President, Yokohama National University
Yokohama, Japan

# Preface

I am very pleased to present this publication, *Sustainable Living with Environmental Risks*, which represents one of the key outputs of the Leadership Programme in Sustainable Living with Environmental Risk (the SLER programme). The Yokohama National University Graduate School of Environment and Information Sciences (YNU-GSEIS) has been implementing the SLER programme over the past few years with the support of the Japanese Ministry of Education, Culture, Sports, Science and Technology (MEXT), and the Japanese Science and Technology Promotion Agency (JST), as well as our partner universities.

To promote trans-disciplinary, pragmatic education and research to facilitate the development of environmental leaders, YNU-GSEIS developed an alliance of faculty members at YNU and partner universities, as well as a curriculum that addresses risk and sustainability from trans-disciplinary, multifaceted perspectives. Every Tuesday, YNU joins with students of the United Nations University-Institute of Advanced Studies (UNU-IAS) and the United Nations Institute of Sustainability and Peace (UNU-ISP) to connect to eight overseas universities through the Interactive Multimedia Education System (iMES). Participants hear lectures given by prominent practitioners working in international research institutes and other organizations, NGOs, business corporations, and government, after which intensive discussions take place.

Our annual SLER intensive course also serves as an exemplar of our participation with partner universities in a wide range of activities. The course includes a tour to observe the process of reconstruction in Rikuzentakata and other areas in Tohoku (northeast Japan) affected by the 2011 disaster. There are also visits to Kawasaki and Hayama, a symposium at the United Nations University, and a scenario workshop at YNU that addresses integrated risk and environment management, exploring the science–policy interface. A three-day workshop on environmental leadership and career development was another unique SLER programme initiative held in Hayama in March 2013 and co-organized with UNU-IAS and ISP.

There are so many more activities that could be cited as achievements of the SLER programme, and this book attests to the extent of our determination to further enhance our education and research programmes for developing environmental leaders. I hope that you will find it both useful and enlightening.

S. Morishita

Shin Morishita
Dean, Yokohama National University
Graduate School of Environment and Information Sciences
Leader, Leadership Programme in Sustainable
Living with Environmental Risk
Yokohama, Japan

# Congratulatory Salutation

As the world population continues to increase exponentially and humans resort to ever more intensified resource use, managing risks and promoting sustainability have become imperative, while the need for environmental leaders is greater than ever.

The Yokohama National University (YNU) Leadership Programme in Sustainable Living with Environmental Risk (the SLER programme) has been supported since 2009 by strategic funding for the promotion of science and technology from the Japanese Ministry of Education, Culture, Sports, Science and Technology (MEXT). The funds were provided as part of the strategic programme for fostering international environmental leaders begun in 2008 to provide support to 17 universities in Japan. The SLER programme is a distinctive one that focuses on environmental risk management and promotion of sustainability, building upon YNU's long-established expertise and achievements in these fields.

I have been associated with several of the key events and activities undertaken in the context of the programme, and therefore had a chance to personally observe this highly productive international partnership among universities. I was especially impressed by its use of the simultaneous broadcasting system for joint lectures and seminars, as well as for symposiums to discuss critical challenges and career development. I am therefore very pleased to recommend this newly launched publication *Sustainable Living with Environmental Risks*, comprising one of the key outputs delivered under the SLER programme.

This textbook contains numerous articles contributed not just by YNU faculty members, but also by faculty in the partner universities. It therefore represents the culmination of all that the programme has achieved. The aim of the textbook is to present up-to-date analysis on environmental risks, their implications, and interventions to promote effective risk management and sustainable development. It reflects insightful views and suggestions for advancing science, education, and leadership development aimed at achieving sustainability.

I would like to congratulate YNU for its successful implementation of the SLER programme and for the publication of this textbook. I am honoured to pay a tribute in this congratulatory salutation on the occasion of the textbook's launch. I sincerely hope that YNU will continue to demonstrate global leadership in research, education, and leadership development, and I am certain that this textbook will inspire its readers to ever greater wisdom and ingenuity in the pursuit of effective risk management and sustainability.

Koujun Yamashita
Program Officer, Department for Promoting Science
and Technology System Reform, Japan Science and Technology Agency

# Acknowledgments

It has been my honour and pleasure to act as Coordinator for the Leadership Development Programme for Sustainable Living with Environmental Risks (SLER). SLER was launched in 2009 and has been being carried out with the support of the Japanese Ministry of Education, Culture, Sports, Science and Technology (MEXT), and the Japanese Science and Technology Promotion Agency (JST). I am grateful to Prof. Kunio Suzuki, Yokohama National University (YNU) and Prof. Shin Morishita, Dean of the YNU Graduate School of Environment and Information Sciences (GSEIS) and Leader of SLER for their ebullient guidance and support.

This book "*Sustainable Living with Environmental Risks*" is a consolidation of knowledge, insights and suggestions for developing environmental leadership on risk management and sustainability promotion deriving from SLER activities carried out over the years. On behalf of the other editors and my colleagues Prof. Shinji Yoshiura and Associate Professor Masanori Kobayashi, I would like to express my gratitude and appreciation particularly to the SLER focal points of partner universities namely Prof. Da Liangjun (East China Normal University), Prof. Ainin Niswati (University of Lampung), Prof. Samuel Kiboi (University of Nairobi), Prof. Bruno Ramamonjisoa (University of Antananarivo), Prof. Norizan Esa and Prof. Nik Norma Nik Hasan (University Science Malaysia), Prof. Roberto F. Rañola Jr. (University of the Philippines Los Baños), Prof. Damrong Pipatwattanakul (Kasetsert University), and Prof. Nguyen Van Dung (University of Danang). I would also like to thank Dr. Koujun Yamashita, Program Officer of the Japan Science and Technology Agency, Prof. Govindan Parayil, Vice Rector of the United Nations University and Director of the United Nations University-Institute of Advanced Studies, Prof. Kazuhiko Takeuchi, Senior Vice-Rector, United Nations University and Director, Institute for Sustainability and Peace, and Mr. Hideyuki Mori, President, Institute for Global Environmental Strategies for their support to SLER as our key collaborating institutes. So many other faculty member colleagues and staff members of YNU and our partner universities and collaborators of our partner institutions have given us tremendous support to SLER and the publication of this book. I ask for your indulgence no to mention all. I hope that this book will be a valuable contribution to our endeavours to develop environmental leaders for a sustainable world.

Nobuhiro Kaneko
Coordinator, Environmental Leadership Programme for Sustainable Living with Environmental Risks
Professor, Yokohama National University Graduate School of Environment and Information Sciences

# Contents

# Chapter 1
# Managing Environmental Risks and Promoting Sustainability, Scientific Advancement, and Leadership Development

**Masanori Kobayashi, Shinji Yoshiura, Takako Sato, and Nobuhiro Kaneko**

**Abstract** As entrenched population growth and industrialization continue to raise demand for natural resources and their exploitation, there is increasing concern over the detrimental impacts on the global environment and humanity. Economic growth was expected to save people from poverty, but conventional economic growth models simply prompted intensive resource use and undermined the basis for livelihoods that are sustainable over the long term. Whilst research and policy measures have articulated environmental risks and key factors of sustainability, compartmentalized approaches have failed to forge a scientific foundation for averting risks and promoting sustainability. Countermeasures to address environmental risks often involve trade-offs weighed against other socio-economic factors. A holistic viewpoint and trans-disciplinary science are therefore needed to foster appropriate decision making and implementation that can ensure optimal risk management and promotion of sustainability. The Leadership Programme in Sustainable Living with Environmental Risk (the SLER programme) spearheaded by Yokohama National University from 2009 to 2014, is one of the programs playing an instrumental role in addressing this need. It provides a platform for strengthening the expertise and skills graduate school students need to become environmental leaders. Moreover, the process of implementing the SLER programme has revealed both the potential and the challenges inherent in developing future environmental leaders to effectively manage environmental risk and promote sustainability.

**Keywords** Environmental risk • Leadership development • Risk trade-off • Sustainability • Trans-disciplinary science

M. Kobayashi (✉) • S. Yoshiura • T. Sato • N. Kaneko
Graduate School of Environment and Information Sciences, Yokohama National University, 79-7 Tokiwadai, Hodogaya-ku, Yokohama City 240-8501, Japan
e-mail: m-kobayashi@ynu.ac.jp; yoshiura@ynu.ac.jp; sugar@ynu.ac.jp; kanekone@ynu.ac.jp

N. Kaneko et al. (eds.), *Sustainable Living with Environmental Risks*,
DOI 10.1007/978-4-431-54804-1_1, 

## 1.1 Introduction: Environmental Risks and Their Implications for Future Sustainability

Economic growth and increased use of resources due to industrialization have raised the pressure on the global environment and serious warnings have been sounded that further pressure could destabilize the Earth's systems and trigger abrupt and irreversible environmental changes (Rockström et al. 2009). The average global surface temperature rose by 0.85 °C over the period 1880–2012, and by 2100 is projected to increase by 2.68–4.8 °C, accompanied by a rise in sea level of up to 0.98 m (IPCC-WGI 2013). The global wild animal population declined by more than 30 % over the period 1970–2010 and the annual economic loss attributable to deforestation and forest degradation could be equal to USD 4.5 trillion (SCBD 2010). Environmental degradation undermines the basis of people's livelihoods and often impoverishes communities. At the same time, poverty drives people to exploit natural resources for their survival and exacerbates environmental degradation in a vicious cycle (Bremner et al. 2010).

In June 2012 global leaders gathered at the United Nations Conference on Sustainable Development held in Rio de Janeiro. In the conference's outcome document, entitled "The Future We Want," they reaffirmed their commitment to promoting sustainable development for our planet and for present and future generations, and to saving the world from poverty and hunger as a matter of urgency (UNGA 2012). In paragraph 259 the document called for countries to strengthen leadership capacity to promote sustainable development and engage citizens and civil society organizations. Among a number of factors enabling sustainable development, the document underlined the importance of supporting educational institutions in (1) conducting research and innovation for sustainable development, and (2) developing high-quality, innovative including the entrepreneurship and professional training required to achieve sustainable development goals.

The Leadership Programme in Sustainable Living with Environmental Risk (the SLER programme) spearheaded by the Yokohama National University Graduate School of Environment and Information Sciences (YNU-GSEIS) is one of a number of programs designed to develop future leaders who will manage risks and promote sustainability. The SLER programme was commenced in 2009 for a 5-year duration with the support of the Japanese Ministry of Education, Culture, Sports, Science and Technology (MEXT) and the Japanese Science and Technology Promotion Agency. The Government of Japan lists environmental science as a priority area in research and development, and promotes innovation by consolidating knowledge and revitalizing research and development capabilities at both universities and private corporations (CSTPJ 2010). It is believed that Japan can make an essential contribution toward achieving sustainable development throughout the world by developing future environmental leaders within its higher education system.

This paper is intended to delineate environmental risks to sustainability, and their characteristics and implications, to examine what the SLER programme's pedagogical approaches and newly invented curriculum have achieved in terms of filling the gap in development of future environmental leaders. It is also designed to provide a forward-looking perspective on how universities can enhance the effectiveness of their programs for developing future environmental leaders.

## 1.2 Environmental Risks, Their Characteristics, and Sustainability Implications

Environmental risks are defined as risks with the potential to fundamentally disrupt the stability of the Earth's systems (IGBP 2012), while risk itself is defined as the combination of the probability of an event and its negative consequences (Nadim 2011). The destabilization of the Earth's systems could trigger environmental changes that would be deleterious or even catastrophic for human beings (Rockström et al. 2009). Such environmental risks encompass a wide range of areas such as climate change, water scarcity, deforestation, land degradation, biodiversity loss, ozone depletion, and chemical pollution. By their nature, environmental risks are characterized by (1) spatial propagation, (2) time-lag occurrence, (3) multiplier effects, (4) accumulation, and (5) irreversibility (Zhang et al. 2010). However, the most striking characteristic of environmental risks is their interconnectedness (IGBP 2012). For instance, excessive logging causes deforestation and destructs wild life habitats, thereby depleting biodiversity. Deforestation not only accelerates soil erosion but prompts the emission of greenhouse gases and their concentration in the atmosphere, thereby increasing the likelihood of climate change and destabilizing the water cycle. A negative change in one of the areas can aggravate another area and vice versa.

Risk management decisions often need to take into account the various trade-offs associated with environmental risks (Power and McCarty 2000). Environmental risks and their countermeasures always entail positive and negative environmental, economic, and social trade-offs (Table 1.1). For instance, rapid reforestation with a newly-introduced species of exotic fast-growing tree may be effective in increasing forest cover and sequestrating carbons, however it may also make the long-term integrity and autonomy of the forest ecosystem uncertain. It would reduce the space for endemic/indigenous tree species and wildlife habitats and would also hinder the access of local villagers to diverse forest resources such as leaves, fodder, fuel woods, and other non-timber products. It is therefore vital to ensure that reforestation would damage neither the environment nor people (Peskett and Todd 2013). We need to safeguard the overall environmental value of forest areas and the interests of local and indigenous people even as we pursue the goals of carbon sequestration, reduction of greenhouse gas emissions, and climate change mitigation (WRI 2012).

Environmental risk trade-offs also need to take into account differing local conditions. DDT (dichloro-diphenyl-trichloroethane) and its application is a classic case often cited to describe environmental risk trade-offs and the complexity involved in assessing and making decisions about such trade-offs (Pfau 2011).

**Table 1.1** Environmental risks and their trade-offs

| Primary environmental risks | Typical countermeasures | + / – | Environmental benefit and trade-off risks | Economic benefit and trade-offs | Social benefit and trade-offs |
|---|---|---|---|---|---|
| Climate change | Promoting biofuel | – | Deforestation, changes in land use | Investment cost | Competition with food |
| | | – | Biodiversity loss | | Loss of access to forest resources |
| | | + | Reduced fossil fuel use | Revenue from the sales of biofuel | Employment opportunities |
| | | + | Reduced GHG emissions | | |
| Biodiversity loss | Increasing protected areas | – | Possible increase in incidents of wild animal attack against humans | Increased demand for budget | Restriction of productive activities<br>Social disturbance by visitors |
| | | – | Possible increase in pests | | |
| | | + | Increased flora and fauna | Possible revenue from park admission fees | Increased employment and income generation opportunities |
| Increased waste generation | Promoting recycling | – | Possible leakage of hazardous substances from recycling plants | Investment in the construction of recycling plants | Need to develop social systems conducive to recycling such as segregated waste collection at source |
| | | + | Reduced demand for resources | Development of recycling businesses | Increased social unity and vigilance |
| | | + | Reduced waste | | |
| Water scarcity | Harvesting rainwater | – | Possible increase in mosquitoes and insects | Investment in installation of micro rain harvesting system | Possible increase in vector disease |
| | | + | Reduced demand for piped water and irrigation | Reduced revenue/increased cost for the water service corporations | Increased water supply sufficiency<br>Reduction in conflict over water use |
| Land degradation | Implementing agroforestry | – | Reduced sunlight on farms | Reduced income from crops | Requirement for coordination with forestry and agriculture groups |
| | | + | Prevention of soil erosion | Diversified income sources | Stability in income |
| | | + | Wind breaking | Averting risks of poor harvests | Respect and self-esteem from innovation and entrepreneurship |
| | | + | Reduced heat and evaporation | Investment in timber production | |
| | | + | Enhanced moisture level | | |
| | | + | Improved nutrient cycle | | |
| | | + | Enhanced biodiversity | | |

Malaria is one of the most lethal diseases in the world. Although the total number of infections is declining gradually, it is estimated that in 2010 there were 219 million cases of infection, of which 79 % occurred in Africa. A total of 660,000 people were killed, with the death toll in Africa accounting for 90 % of these (WHO 2012). DDT is considered to be the most cost-effective insecticide for containing malaria (Pedercini et al. 2011). However, it is known that DDT may have a variety of human health effects, including reduced fertility, genital birth defects, breast cancer, diabetes, and damage to developing brains. In addition, its metabolite DDE (dichloro-diphenyl-dichloroethylene) can block male hormones (Cone 2009). DDT's stigma was made known to the world by Rachel Carson's "Silent Spring," published in 1962 (Dugger 2006). DDT and DDE stay in the environment long-term and their bio-magnification threatens animals at higher trophic levels. Despite being banned in many countries during the 1970s on the grounds of its adverse effect on human health and ecosystems, DDT has been used particularly in developing countries to control malaria (Secretariat of the Stockholm Convention 2013). The Stockholm Convention on Persistent Organic Pollutants, adopted in 2001 and enforced in 2004, lists DDT as one of the "persistent organic pollutants" to be banned or regulated. On the other hand, in 2006, the World Health Organisation (WHO) reversed nearly 30 years of policies restraining the use of DDT and instead endorsed DDT use for indoor residual spraying (IRS) in epidemic areas as well as in areas with constant and high malaria transmission (WHO 2006; Boddy-Evans 2006).

As people have different perceptions of malaria risk, the use of DDT remained contentious, while associated measures to tackle malaria were carried out in ways that outraged communities. In one case in Uganda the government decided to start spraying, but did not give any advance warning to the communities, let alone consulting with them beforehand. Houses were sprayed even when people were not at home and food and cotton harvests had been left exposed. People were complaining that after the DDT spraying women suffered miscarriages and cattle died, but those who refused DDT spraying were imprisoned. Meanwhile, their cotton produce was rejected in the ecological market on the grounds of marginal DDT traces. It was rumored that corruption between the government and the chemical industry was involved, and that malaria risks had been exaggerated, and false claims made that alternatives to DDT were unavailable (Den Berg 2010). In fact, alternatives to DDT were promoted in a global program launched by the Global Environment Facility, WHO, and the United Nations Environment Programme in 2008. The program advocated integrated vector management including use of a mosquito-net, repellent, and mosquito coils (UNEP and WHO 2008).

The example of malaria risk management reveals the variability and complexity involved. Clearly, risk management must move beyond the assessment of a single risk to mobilize multi-disciplinary expertise in assessing multiple scientific and social risks (Pfau 2011) (Fig. 1.1). Moreover, stakeholder involvement is pivotal in developing and implementing long-term and self-reliant measures for managing risks and promoting sustainability. Public access to information, communication of risks, and stakeholder participation in decision-making are all fundamental to the process of determining countermeasures.

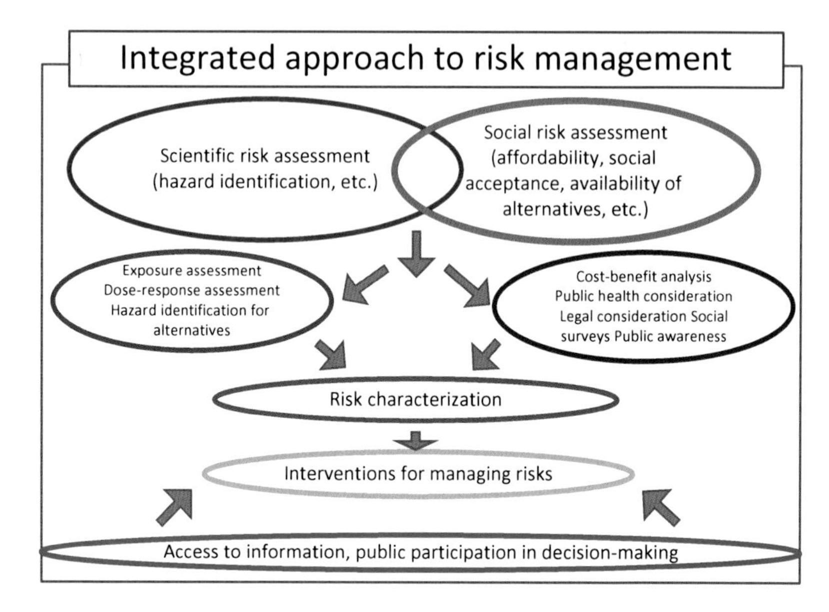


**Fig. 1.1** Integrated approach to risk management (developed from Pfau 2011)

Effective measures for managing risks and promoting sustainability call for trans-disciplinary, multi-partnership, multi-dimensional research (Dedeurwaerdere 2013; Earth System Governance Project (ESGP) 2012). Spearheaded by the International Council for Science (ICSU) and others, the newly launched Future Earth sustainability research initiative is expected to play a key role in providing a reinforced, overarching framework for sustainability science (Yasunari 2013). The platform for enhancing the science-policy interface needs to be bolstered by building upon the prototype recently provided by the Intergovernmental Platform on Biodiversity and Ecosystem Services (Takeuchi 2013). Sciences that address risks and sustainability are changing to involve multiple actors in addressing issues from a trans-disciplinary perspective across a wider range of temporal and spatial scales (Benn et al. 2008). Universities must be pivotal players in transforming the platform of trans-disciplinary science to support risk management and the promotion of sustainability.

## 1.3 Developing Expertise and Skills for Future Environmental Leaders

An increasing number of initiatives and programs have been launched by universities, NGOs, business communities, and research institutes to develop more environmental leaders who can contribute to building sustainable societies. These programs are essentially designed to help students or participants develop (1) scientific expertise, (2) the ability to plan solutions, and (3) skills to steer the implementation process. They also provide a platform for dialogues with leading practitioners (MOEJ 2011;

CLiGS 2013; Wharton-UPENN 2013). Current understanding of leadership and its relation with environment and sustainability is in the developmental stage and it is inevitable that such understanding will evolve over time (Redekop 2010). Heifetz et al. (2009) assert that leadership needs to be adaptive, and adaptive leadership is crucial for thriving on experimentation and mobilizing people to tackle tough and varying challenges. Various attempts have been made to define environmental leadership or leadership for sustainability, and they can be summed up in the phrase "the ability to mobilize and direct people toward achieving sustainability in a changing world."

This then leads us to ask what capabilities can be developed to enable people to play a role as environmental leaders. Williams (2010) presents various skills and expertise that qualify people as environmental leaders, such as (1) technical knowledge, (2) facilitation skills, (3) direction setting, (4) securing resources, (5) creativity, (6) developing relationships, (7) making decisions, (8) communication, (9) determination, and (10) mentoring. Thomas (1993) underscores the personal and professional ethics involved in the leadership role in terms of, for example, complying with rules and norms, and putting the public interest first. However, sustainability issues and their management have become so complex that policies and laws cannot necessarily articulate every detail, and the behavior of practitioners can therefore vary. Moral choice constitutes a critical issue, particularly when practitioners encounter situations for which there are no preceding governing norms. It is therefore still a challenge to know how to address ethical, moral, and value judgments in leadership development.

At the Joint Congress of Environmental Leaders Program 2013, the Japanese universities that are currently implementing, or have already implemented, environmental leadership programs at the graduate school level presented their progress and outcomes (Tsukuba University 2013). Representatives of 17 universities in Japan gave presentations, many of which highlighted the features and characteristics of their particular programs. The common elements are summarized as follows: (1) English language-based, involving non-Japanese students and teaching staff, (2) a cross-sectoral approach addressing the nexus of various interwoven environmental and sustainability issues, (3) an inter-disciplinary curriculum requiring students to learn disciplines other than their major, (4) regional and global features to train students in thinking beyond national borders, (5) development of pragmatic skills, such as communication, writing, and facilitation, (6) internship, (7) partnership with other Japanese and overseas universities, and (8) dialogues with practitioners. The programs are operated primarily with funding from the Japanese Ministry of Education, Culture, Sports, Science and Technology (MEXT), and the Japanese Science and Technology Promotion Agency (JST) for a 5-year duration. Five universities completed their 5-year programs in March 2013. Seven others will complete their programs in March 2014, to be followed by the other five in March 2015. By and large, all the programs are considered to have performed satisfactorily in achieving their stated objective of promoting sustainability science and leadership development in universities.

Nonetheless, it remains a challenge to measure the effectiveness and impacts of such environmental leadership development programs. The universities can of course cite how many students completed their programs and obtained master's or doctoral degrees, and the sectors in which they were employed after graduation. Indeed, such figures are useful indicators of the programs' achievements. However, it will take some time to find out what role the program alumni eventually play as environmental leaders.

The faculty members of universities responsible for the programs are especially concerned about the continuation of the programs, and the associated institutional set-up. All the universities operate under stringent budgets and depend on external resources provided as subsidies by MEXT and JST. Budgets for operating the programs are not yet integrated into universities' core budgets, and it is unlikely that they ever will be. In the past, some universities received subsidies for different but related projects, which took over at least some of the activities in the environmental leadership development program. Many program coordinators in the universities with programs under way or approaching their conclusion are experiencing difficulty in arranging for their programs to be integrated into operations funded by their university's core budget once the 5-year funding by MEXT and JST ceases. It therefore remains to be seen over the coming years where future environmental leadership development programs will take place and how they will evolve. It is undoubtedly a challenge for many universities to find a way of integrating these programs' activities into operations funded by core budgets, or to secure alternative sources of funding for their continuation.

## 1.4 Leadership Programme in Sustainable Living with Environmental Risk

The Leadership Programme in Sustainable Living with Environmental Risk (the SLER programme) was launched in 2009 and is spearheaded by the Yokohama National University Graduate School of Environment and Information Sciences (YNU-GSEIS). The SLER programme has its own distinctive features aimed at developing the expertise and skills required for future environmental leaders (Kaneko et al. 2013). Many features are similar to those implemented by other universities as presented in the previous section. Some of the key features of the YNU-SLER programme are highlighted below (Fig. 1.2).

### *1.4.1 Interactive Multimedia Education System (iMES) (Arisawa and Sato in This Book)*

YNU collaborates with nine overseas universities, namely: East China Normal University (ECNU, China); University of Lampung (UNILA, Indonesia); Universiti Sains Malaysia (USM, Malaysia); University of the Philippines Los Baños (UPLB, Philippines); Kasetsart University (KU, Thailand); The University of

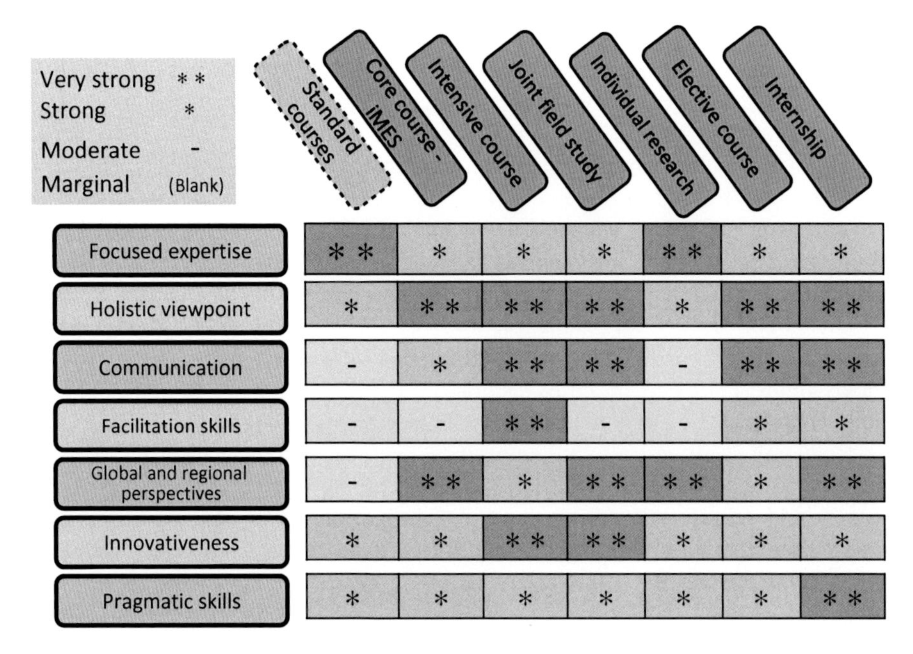


**Fig. 1.2** SLER programme components and their expected impacts in development of expertise and skills required for future environmental leaders

Danang (UOD, Vietnam); University of Nairobi (UON, Kenya); the University of Antananarivo (UOA, Madagascar), and the United Nations University (UNU). The students of the UNU Institute of Advanced Studies (UNU-IAS, Yokohama) attend the class at YNU, and students from the UNU Institute of Sustainability and Peace (UNU-ISP, Tokyo) are now invited to do so from fall 2013. The two core courses of the SLER programme are Environmental Risk Management (spring semester) and Environmental Leadership Development (fall semester), and they use the interactive multimedia education system (iMES). Lectures are given in English using PowerPoint for 25 min by guest speakers from international organizations, research institutes, NGOs, business corporations, and governments. The moderators based at YNU facilitate discussions involving both the YNU students and students in overseas universities connected via iMES.

### *1.4.2 Intensive Course*

In September each year, a 2-week intensive SLER course is organized with participants from YNU, UNU, and eight overseas partner universities. The program includes some unique components, notably (1) a tour to study reconstruction in the parts of Tohoku (northeastern Japan) hit by the 2011 earthquake and tsunami disaster, (2) a visit to the city of Kawasaki near Tokyo to learn about the operation of environmental businesses by both private and public organizations, (3) a visit to the town of Hayama on Tokyo Bay's Miura Peninsula to learn integrated ecosystem and landscape management from local people, (4) dialogues with leading scientists and practitioners at UNU's open joint symposium, and (5) a scenario workshop to develop students' creative thinking and facilitation skills. During the study tour to Tohoku in September 2013, the students observed and interacted with experts and local stakeholders in the city of Nihonmatsu in Fukushima Prefecture, the town of Minamisanriku and the city of Iwanuma in Miyagi Prefecture, and the city of Rikuzentakata in Iwate Prefecture. They learned about (1) the grand design for reconstruction, (2) forest management in radiation-affected areas, (3) coastal woodland restoration, (4) restoration of tsunami-inundated paddy fields, (5) oyster farming restoration, (6) debris and waste management, and (7) revitalization of small and medium enterprises. Students also participated in producing seedling pots containing local evergreen broad-leaved tree species such as laurel or *persia thumbertii*. The YNU Student Association of Films produced a 45-minute video featuring the Tohoku study tour of 2011–2012 that was screened at a public symposium entitled Reconstruction and Invigoration of Disaster-hit Areas—Viewpoints from Rikuzentakata, held at YNU on March 25, 2013.

### 1.4.3 Madagascar Joint Field Study

A joint field study is conducted once a year in Madagascar in collaboration with UOA. In 2012, YNU and UOA students undertook (1) an ecosystem assessment, (2) a soil survey, and (3) a social survey in the areas of Ambatondrazaka, on Madagascar's eastern side, and Andapa further to the north-east. Together with experts and local practitioners students observed and discussed: (1) environment, forest, and agriculture policy issues, (2) management of protected and watershed areas, (3) reforestation, (4) non-tillage farming, (5) wildlife protection, and (6) innovative community-based activities offering alternative livelihoods. A joint symposium was organized at UOA on the last day of the program to present the outcomes of the field study.

### 1.4.4 Credit Exchange Agreement with UNU

In order to formalize its academic and educational collaboration with UNU-IAS, YNU-GSEIS entered into a credit exchange agreement with UNU-IAS in March 2012. Between then and July 2013, four students of YNU took three courses at UNU-IAS, and nine students of UNU-IAS took three courses at YNU. The credit exchange agreement was expanded in July 2013 with the conclusion of an additional agreement among YNU, UNU-IAS, and UNU-ISP. With the addition of UNU-ISP, the four other YNU graduate schools have now joined in the agreement.

### 1.4.5 *Other Elective Courses and Supporting Programs*

The Asia-Africa Field Work II course helps students to conduct their individual field surveys. Effective Communication for Environmental Leaders enables students to improve their writing and communication skills. Additional elective courses offered in English are Local Risk and Resource Management, International Cooperation for Sustainable Development, and Eco-tourism. A course entitled Capacity Development that was left at the conceptual stage for some time was operationalized in 2012 to give credits when the students undertake internships or attend seminars overseas. In March 2012 an ad-hoc two-day seminar entitled Workshop for Environmental Leadership and Career Development was held in Hayama. Organized in collaboration with UNU-IAS and UNU-ISP, the workshop featured 16 speakers from international organizations, international NGOs, and business corporations.

## 1.5 Achievements and Future Challenges

During the 5-year period from 2009 to 2013, a total of 257 students enrolled in the SLER programme (64 students for the 2–3 year long-term course at YNU, and 193 students of overseas partner universities for the 1 year short-term course via the simultaneous broadcasting system). As of October 2013, 91 students had already completed their courses (27 for the long course, 64 for the short course). By March 2014, an additional 110 students are expected to have completed the SLER programme (47 for the long course and 63 for the short course). These statistics far exceed the original targets set out in 2008. Moreover, the administration of the SLER programme is highly acclaimed by and large, with the students earning awards in poster competitions and acknowledging the support provided to them. At the same time, however, implementation of the program revealed some future challenges with regard to supporting environment/sustainability leadership development:

### 1.5.1 *Curriculum Development*

The SLER programme is expected to generate spin-off courses in English addressing environmental risks and sustainability, and the five YNU graduate schools are expected to align themselves more proactively in support of the program. There is, however, a need to provide further stimulative measures for creating the desired curriculum and ensuring the necessary institutional evolution.

### 1.5.2 Institutional Set-up

The SLER administrators are proposing to establish an international center for risk studies. A number of options were discussed in the past, including restructuring the existing YNU Center for Risk Management and Safety Sciences (CRMSS) or creating a new center to strengthen research and education on environmental risk and sustainability issues. Although some CRMSS personnel work on environmental risk and sustainability issues, however, the center's portfolio is currently focused on industrial engineering, construction infrastructure, urban disaster management, chemical risks, and road traffic safety. Furthermore, the Japanese name for CRMSS is different from the English, describing CRMSS as a center for "safety and security," which may not be helpful in accessing funds or developing partnerships on environmental risk- or sustainability-related issues. Further institutional changes are therefore required to forge a structure for following up and capitalizing on the SLER programme.

### 1.5.3 Institutionalizing Collaborative Educational Activities

A proposal was initially considered to expand credit exchange agreements and to introduce a double degree program among YNU and overseas partner universities. Preliminary discussions were held on the requirements and advantages, but no clear decision was forthcoming due to unresolved practical issues such as consistency with the respective universities' existing degree requirements, and differences in academic calendars.

### 1.5.4 iMES

iMES has been recognized as an extremely useful system for collaboration among universities in risk management and sustainability research and education. However, the staff members who operate the system currently depend on non-budget subsidies, and it is not yet clear whether and to what extent existing organizations such as the YNU Information Technology Service Center can support continuation of the course jointly conducted with other universities.

### 1.5.5 Joint Research

There have been a number of calls to promote joint research between YNU and one or more of its partner universities. While there was ongoing collaborative research predating the start of the SLER programme, and some attempts have been made to launch new projects, there has not yet been any success in mobilizing funds to launch new joint research initiatives.

The SLER programme helped to generate future environmental leaders and set a very useful platform and roadmap for invigorating environmental leadership development among YNU and its partner universities, as well as associated organizations, experts, and stakeholders. There are still ways to move forward more vigorously and expeditiously to reduce, halt, and reverse the accelerating environmental risks and to bolster partnership and collective action for a sustainable world. The spirit and compassion generated through the SLER programme must not be allowed to dissipate, but must instead be sustained and bolstered.



## References

Benn S, Dunphy D, Martin A (2008) Governance of environmental risk: new approaches to managing stakeholder involvement. J Environ Manage 90:1567–1575

Boddy-Evans A (2006) World Health Organization okays DDT. African History. http://africanhistory.about.com/b/2006/09/16/world-health-organization-okays-ddt.htm. Accessed 21 Sept 2013

Bremner J, López-Carr D, Suter L, Davis J (2010) Population, poverty, environment, and climate dynamics in the developing world. Interdiscipl Environ Rev 11(2/3):112–126, http://geog.ucsb.edu/~carr/wordpress/wp-content/uploads/2012/04/Bremneretal_IntEnvRev_201012.pdf

Center for Leadership in Global Sustainability (CLiGS) (2013) Achieving the mission: strategic opportunity through partnership. 23 July 2013. http://cligs.vt.edu/short-course-registration-now-open/. Accessed 23 Sept 2013

Cone M (2009) DDT use should be last resort in malaria-plagued areas, scientists say. Environmental Health News. 4 May 2009 http://www.environmentalhealthnews.org/ehs/news/ddt-only-as-last-resort. Accessed 21 Sept 2013

Council for Science and Technology Policy of Japan (CSTPJ) (2010) Japan's science and technology basic policy report. http://www8.cao.go.jp/cstp/english/basic/4th-BasicPolicy.pdf. Accessed 17 Sept 2013

Dedeurwaerdere T (2013) Transdisciplinary sustainability science at higher education institutions: science policy tools for incremental institutional change. Sustainability 5:3783–3801

Den Berg JV (2010) It's our turn to eat. Silent Snow. http://www.silentsnow.org/nl/121. Accessed 22 Sept 2013

Dugger CW (2006) W.H.O. supports wider use of DDT vs. malaria. New York Times. 16 September 2006. http://www.nytimes.com/2006/09/16/world/africa/16malaria.html?_r=2&. Accessed 21 Sept 2013

Earth System Governance Project (ESGP) (2012) Rio+20 Policy Brief Transforming governance and institutions for a planet under pressure. http://www.icsu.org/rio20/policy-briefs/InstFrameLowRes.pdf. Accessed 17 Sept 2013

Heifetz RD, Grashow A, Linsky M (2009) The practice of adaptive leadership – tools and tactics for changing your organization and the world. Harvard Business, Boston

Intergovernmental Panel on Climate Change Working Group I (IPCC-WGI) (2013) Summary for policymakers of the working group I contribution to the IPCC Fifth Assessment Report. http://www.climatechange2013.org/images/uploads/WGIAR5-SPM_Approved27Sep2013.pdf. Accessed 28 Sept 2013

International Geosphere-Biosphere Programme (IGBP) (2012) Rio+20 Policy Brief Interconnected risks and solutions for a planet under pressure. http://www.icsu.org/rio20/policy-briefs/interconnected-issues-brief. Accessed 17 Sept 2013

Kaneko N, Yoshiura S, Kobayashi M, Sato T (2013) Yokohama National University Leadership Development Programme for sustainable living with environmental risks. In: Tsukuba University (ed) Proceedings of the joint congress of environmental leaders program 2013. Tsukuba University, Tokyo, pp 10–13

Ministry of Environment, Japan (MOEJ) (2011) Vision for environmental leadership initiatives for Asian sustainability. https://edu.env.go.jp/asia/en/about/vision.html. Accessed 23 Sept 2013

Nadim F (2011) Risk, hazard and vulnerability. Risk assessment and mitigation training workshop. https://www.gfdrr.org/sites/gfdrr.org/files/01_Hazard_and%20_RiskTerminology.pdf. Accessed 17 Sept 2013

Pedercini M, Blanco SM, Kopainsky B (2011) Application of the malaria management model to the analysis of costs and benefits of DDT versus non-DDT malaria control. PLoS One 6(11):e27771, http://www.plosone.org/article/info:doi/10.1371/journal.pone.0027771. Accessed 20 Sept 2013

Peskett L, Todd K (2013) Putting REDD+ safeguards and safeguard information systems into practice. The United Nations Collaborative Programme on Reducing Emissions from Deforestation and Forest Degradation in Developing Countries (UN-REDD) Policy Brief No.3. http://www.un-redd.org/Newsletter35/PolicyBriefonREDDSafeguards/tabid/105808/Default.aspx. Accessed 19 Sept 2013

Pfau K (2011) A risk-risk trade-off: Insecticide use for malaria control. http://dukespace.lib.duke.edu/dspace/handle/10161/3678. Accessed 19 Sept 2013

Power M, McCarty LS (2000) Risk-cost trade-offs in environmental risk management decision-making. Environ Sci Policy 3:31–38

Redekop BW (2010) Connecting leadership and sustainability. In: Redekop BW (ed) Leadership for environmental sustainability. Leadership for Environmental Sustainability, New York, pp 1–15

Rockström J, Steffen W, Noone K, Persson A, Chapin FS III, Lambin E, Lenton TM, Scheffer M, Folke C, Schellnhuber H, Nykvist B, De Wit CA, Hughes T, van der Leeuw S, Rodhe H, Sölin S, Snyder PK, Costanza R, Svedin U, Falkenmark M, Karlberg L, Corell RW, Fabry VJ, Hansen J, Walker B, Liverman D, Richardson K, Crutzen P, Foley J (2009) Planetary boundaries: exploring the safe operating space for humanity. Ecol Soc 14(2):32, http://www.ecologyandsociety.org/vol14/iss2/art32/. Accessed 20 Sept 2013

Secretariat of the Convention on Biological Diversity (SCBD) (2010) Global Biodiversity Outlook 3. http://www.cbd.int/gbo3/. Accessed 15 Sept 2013

Secretariat of the Stockholm Convention on Persistent Organic Pollutants (2013) The 12 initial POPs under the Stockholm Convention. http://chm.pops.int/TheConvention/ThePOPs/The12InitialPOPs/tabid/296/Default.aspx. Accessed 20 Sept 2013

Takeuchi K (2013) Science and innovation for a sustainable future: the role of sustainability science. International Symposium "Developing leaders for managing risks and promoting sustainability toward establishing a resilient and sustainable society." Tokyo. http://www.sler.ynu.ac.jp/node/592. Accessed 23 Sept 2013

Thomas JW (1993) Ethics for leaders. In: Berry JK, Gordon JC (eds) Environmental leadership – developing effective skills and styles. Island, Washington D.C, pp 31–45

Tsukuba University (2013) Proceedings of the Joint Congress of environmental leaders program 2013. Tokyo, 14 September 2013. Tsukuba University

United Nations Environment Programme (UNEP) and World Health Organization (WHO) (2008) Towards DDT-free malaria control http://www.unep.org/PDF/UNEP_GEFMalariaLeaflet.pdf. Accessed 22 Sept 2013

United Nations General Assembly (UNGA). 66/2888. The future we want. 11 September 2012. http://www.un.org/en/ga/search/view_doc.asp?symbol= A/RES/66/288. Accessed 16 Sept 2013

Wharton School, University of Pennsylvania (Wharton-UPENN) (2013) Initiative for global environmental leadership sixth annual conference-workshop. The nexus of energy, food and water. http://lgstdept.wharton.upenn.edu/igel/Conf2013_Brochure.pdf Accessed 24 Sept 2013

Williams RL (2010) Leadership and the dynamics of collaboration – averting the tragedy of the commons. In: Redekop BW (ed) Leadership for environmental sustainability. Leadership for Environmental Sustainability, New York, pp 67–92

World Health Organization (WHO) (2006) WHO gives indoor use of DDT a clean bill of health for controlling malaria. http://www.who.int/mediacentre/news/releases/2006/pr50/en/. Accessed 20 Sept 2013

World Health Organization (WHO) (2012) World malaria report 2012 http://www.who.int/malaria/publications/world_malaria_report_2012/report/en/index.html. Accessed 19 Sept 2013

World Resource Institute (WRI) (2012) Safeguarding forests and people: a framework for designing a national system to implement REDD+ safeguards. http://pdf.wri.org/safeguarding_forests_and_people.pdf. Accessed 18 Sept 2013

Yasunari T (2013) Future earth – a new international initiative toward global sustainability. International Symposium "Developing leaders for managing risks and promoting sustainability toward establishing a resilient and sustainable society." Tokyo. http://www.sler.ynu.ac.jp/node/592. Accessed 23 Sept 2013

Zhang K, Pei Y, Lin C (2010) An investigation of correlations between different environmental assessments and risk assessment. Procedia Environ Sci 2:643–649

# Part I
# Sustainable Primary Production for Human Well-Being

# Chapter 2
# Biodiversity Agriculture Supports Human Populations

**Nobuhiro Kaneko**

**Abstract** The "Green Revolution" has increased food production to meet world population growth, therefore global food production is at present sufficient to feed all the world's people. However, the modern agricultural system is no longer sustainable due to deterioration of soil conditions. Alternative agricultural methods that aim to conserve biodiversity and soil functioning are not intensively studied, thus the productivity of alternative methods is often not compatible with conventional agricultural practice, and most people are skeptical of the feasibility of introducing alternative methods. Recent advancements in studies of biodiversity and ecological functioning are now supporting early trials by advanced farmers, who respect biodiversity in their fields. In this review, I would like to present some ecological theories to support biodiversity agriculture and its potential to support human populations.

**Keywords** Above- and below-ground interaction • Conservation tillage • Ecological theory • Food security • Soil conservation

## 2.1 Introduction

The world produces sufficient food to feed its population, but still there remain more than one billion people who suffer from food insecurity and malnutrition (IAASTD 2009). Agricultural activities are now one of the major factors affecting global environmental change through reducing biodiversity, increasing greenhouse gas (GHG: $CO_2$, $N_2O$, and $CH_4$) emissions and accelerating eutrophication and pollution of aquatic systems. Agriculture's main challenge will be to produce sufficient food and fiber for a growing global population at an acceptable environmental cost.

N. Kaneko (✉)
Graduate School of Environment and Information Sciences,
Yokohama National University, Yokohama, Japan
e-mail: kanekono@ynu.ac.jp

N. Kaneko et al. (eds.), *Sustainable Living with Environmental Risks*,
DOI 10.1007/978-4-431-54804-1_2, © The Author(s) 2014

To meet the global demand for food without significant increases in prices, it has been estimated that we need to produce 50–110 % more food, in light of the growing impacts of agricultural activities on climate change (Godfray et al. 2010; Tilman et al. 2011). Closing 'yield gaps' on underperforming lands, increasing cropping efficiency, shifting diets, and reducing waste can double food production while greatly reducing the environmental impacts of agriculture (Foley et al. 2011). However, most of these discussions are based on the assumption that the agricultural production is sustainable using modern agricultural technology.

Biodiversity loss is obviously a major driving force of ecosystem change (Hooper et al. 2012). Even in the field of agriculture, loss of biodiversity is linked to degradation of ecosystem function. Carbon loss (decomposition), nitrogen mineralization, and leaching are influenced by both land use and soil biota (de Vries et al. 2013). Intensification of agriculture decreases soil biodiversity, thus soil degradation is inevitable under the modern agricultural system. Soil degradation includes physical factors (e.g., decline in soil structure, crusting, compaction, accelerated erosion); chemical factors (e.g., nutrient depletion, elemental imbalance, acidification, salinization); and biological factors (e.g., reduction in soil organic matter (SOM), and the activity and species diversity of soil microorganisms) (Lal 2004). Conserving soil biodiversity and utilizing its ecosystem functioning is beneficial not only for agriculture but also for consumers (Robertson and Swinton 2005). Many ecosystem services are synergistic; for example, soil carbon storage keeps $CO_2$ from the atmosphere and also promotes soil fertility, soil invertebrate diversity, plant water-use efficiency, and soil conservation (Lal 2004), and these ecosystem services are supported by biodiversity (Hooper et al. 2005)

## 2.2 Green Revolution and Organic Farming

Green Revolution technology has been criticized for its deficiencies (Swaminathan 2006; Robertson and Swinton 2005). Economists stress that, because market-purchased inputs are needed for production, only resource-rich farmers can take advantage of high-yielding crops. Environmentalists emphasize that the excessive use of fertilizers and pesticides, as well as the monoculture of a few crop cultivars, will create serious environmental problems, including the breakdown of resistance in plants and the degradation of soil fertility by disturbing the stoichiometry.

Organic farming is probably an alternative to modern intensive farming. Organic farming uses organic fertilizers to sustain fertility of soil and reduce chemicals to control pests and weeds. Badgley et al. (2007) argued that the global population can be supported by organic farming. The principal objection to the proposition that organic agriculture can contribute significantly to the global food supply is the fact that organic yields are typically lower than conventional yields (Seufert et al. 2012), necessitating more land to produce the same amount of food as conventional farms. However, when using best organic management practices yields are closer to conventional yields (−13 %). Organic agriculture also performs better under certain conditions—for example, organic legumes or perennials, on weak-acidic to weak-alkaline soils, in rainfed conditions, achieve yields that are only 5 % lower than conventional yields (Seufert et al. 2012).

## 2.3 Biodiversity, Ecological Functioning, and Ecosystem Services

Recent experimental advancements in biodiversity and ecological functioning were mainly obtained in laboratory microcosms, benthos communities (Cardinale et al. 2012), and model grassland studies (Tilman et al. 2001). Varying component species numbers from one (monoculture) to 16 species (average species richness in natural grasslands in Minnesota), Tilman et al. (2001) showed that plant production was higher with increasing species richness. Increase in species richness heightens the chances of involving functional species (sampling effect; c.a. legumes that fix atmospheric nitrogen thus increase nitrogen resources in soil) and also enhances functional diversity (niche complementarity effect).

Niche complementarity explains the decrease in nitrate nitrogen concentration in soil with increasing plant species richness. The greater the species richness, the more efficiently the plant uptakes nitrogen by root due to complement root depth and morphology compared to monoculture soil.

The longer the experiment continues, the higher the stability of primary production (Tilman et al. 2006b). This is explained by the portfolio effect of different species that respond differently to environmental conditions such as drought, high and low temperature, etc. The ratio of predators to prey in aboveground communities tends to increase with plant species richness (Haddad et al. 2011). Therefore this mechanism also contributes to the stability of primary production. Tilman et al. (2006a) concluded that natural short grass prairie is the most productive for supplying biomass for energy use under no-fertilization conditions.

Organic farming avoids utilization of synthetic fertilizers, chemical pesticides, and herbicides. The recent ecological studies on biodiversity suggest that increasing plant species richness leads to efficient nutrient use, fewer outbreaks of pests and pathogens, and stable yields for a certain period. Some organic farming techniques, such as intercropping, use of companion plants, patchy land use, and also agroforestry all increase plant biodiversity compared to monoculture.

## 2.4 Soil Sustainability

Modern agricultural practice is degrading soil by accelerating erosion (Montgomery 2004) and disturbing stoichiometry of nutrients essential for human health (Jones et al. 2013). Organic farming is a system aimed at producing food with minimal harm to ecosystems, animals, or humans. Intensive cultivation, on the other hand, always damages soil biodiversity. Not surprisingly, therefore, comparisons of soil biodiversity in conventional and organic farming showed that conventional farming had been more damaging (Altieri 1999; Chappell and LaValle 2009; Gomiero et al. 2011). Conservation tillage offers an alternative approach involving soil management practices that minimize the disruption of the soil's structure, composition, and natural biodiversity, thereby minimizing erosion and degradation, and water contamination (Holland 2004).

Soil carbon is a good indicator of soil functions. The following technologies should be considered to increase C sequestration in cropland soil: no- and reduced-tillage; residue mulch and cover crops; integrated nutrient management; and biochar used in conjunction with improved crops and cropping systems (Lal 2009). However, biodiversity both above- and belowground also plays an important role in increasing soil carbon. High-diversity mixtures of perennial grassland plant species stored 500 % and 600 % more soil C and N than, on average, did monoculture plots of the same species during a 12-year-long grassland experiment (Fornara and Tilman 2008).

Soil aggregate stability depends on plant community properties, such as composition of functional groups, diversity, and biomass production. Soil aggregate stability increased significantly from monocultures to plant species mixtures (Pérès et al. 2013). Root-derived carbon (C) is preferentially retained in soil compared to aboveground C inputs, and microbial communities assimilating rhizodeposit-C are sensitive to their microenvironment. There was ten times more labeled microbial-C (derived from living roots) in the rhizosphere compared to non-rhizosphere soil (Kong and Six 2012). Weeds are considered to be useless in agricultural soil. However, keeping living roots, even of weeds, can enhance microbial activities in soil.

## 2.5 Soil Biodiversity and Its Functioning

Producing enough food with fewer effects on the environment requires a radical shift in thinking by the agricultural and environmental communities.

Plants are competing neighbors in terms of both shoots and roots. Wild plants are stronger competitors for below-ground resources than are crop plants (Kiaer et al. 2013). However, some Japanese advanced farmers have been successfully growing crops with weeds in their croplands. Introducing weeds as living mulch increases plant biodiversity including crop species. The farmers reduced tillage and fertilization, and no chemical pesticides are sprayed. All these practices enhance not only the aboveground biodiversity of both plants and insects but also soil biodiversity due to less soil disturbance, and increase the amount and diversity of resources for soil organisms. These farmers slash weeds several times during the crop-growing season, thereby reducing aboveground competition between weeds and crops. Adopting these practices is expected to increase soil biodiversity.

Experimental introduction of no-tillage with weed management rapidly changed the soil microbial community and soil carbon sequestration. Our trial in Sumatra,

Indonesia, compared tillage, and no-tillage with a combination of bagasse mulch, in a sugarcane plantation, where productivity of sugarcane had been declining during 30-years of continuous cropping after clear-cutting the forest. All weeds at tillage treatments were suppressed by herbicide, whereas those at no-tillage treatments were hand picked. Soil fungal community structure clearly reflected the treatment one-year after beginning the experiment, and there was an increase in soil carbon content (Miura et al. 2013).

Converting from conventional practice to no-tillage with weed mulch efficiently increased the soil carbon pool despite the carbon input to the soil being very small compared to standard manure application in Japan (Arai et al. 2014).

## 2.6 Conclusion

Comparison of conventional practice and no-tillage with weed mulch shows a contrast, especially in terms of soil biodiversity (Fig. 2.1). There is a growing body of theoretical and empirical studies on the capacity of biodiversity within cropland to improve both stability in production and synergetic effects on production, and to reduce crop loss due to pests and pathogens. These studies are suggesting that ecological analysis is urgently needed to support a novel sustainable cropping system that will support and be supported by biodiversity other than crop plants.

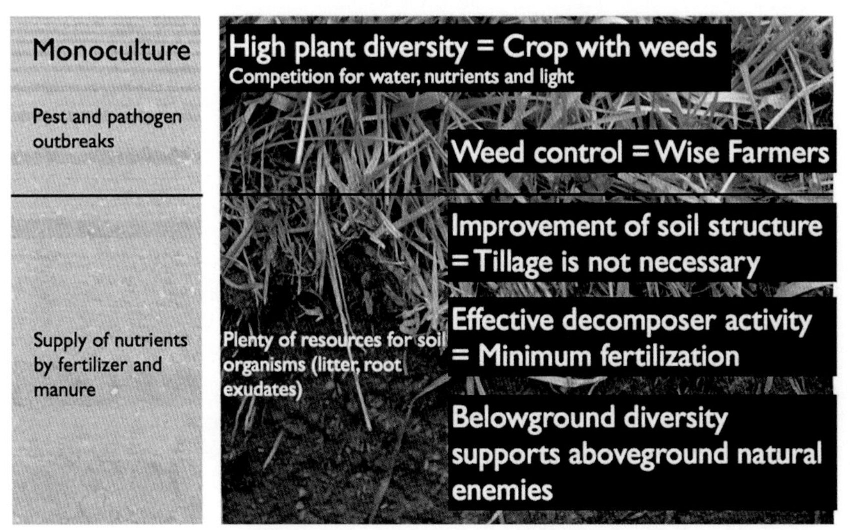


**Fig. 2.1** Conventional farming and farming using no-tillage with weed mulch

## References

Altieri MA (1999) The ecological role of biodiversity in agroecosystems. Agr Ecosyst Environ 74(1–3):19–31. doi:10.1016/S0167-8809(99)00028-6

Arai M, Minamiya Y, Tsuzura H, Watanabe Y (2014) Changes in soil carbon accumulation and soil structure in the no-tillage management after conversion from conventional managements. Geoderma (under review)

Badgley C, Moghtader J, Quintero E, Zakem E, Chappell MJ, Avilés-Vázquez K, Samulon A, Perfecto I (2007) Organic agriculture and the global food supply. Renew Agr Food Syst 22:86. doi:10.1017/S1742170507001640

Cardinale BJ, Duffy JE, Gonzalez A, Hooper DU, Perrings C, Venail P, Narwani A, Mace GM, Tilman D, Wardle DA, Kinzig AP, Daily GC, Loreau M, Grace JB, Larigauderie A, Srivastava DS, Naeem S (2012) Biodiversity loss and its impact on humanity. Nature 486(7401):59–67. doi:10.1038/nature11148

Chappell MJ, LaValle LA (2009) Food security and biodiversity: can we have both? An agroecological analysis. Agric Human Values 28:3–26. doi:10.1007/s10460-009-9251-4

de Vries FT, Thebault E, Liiri M, Birkhofer K, Tsiafouli MA, Bjornlund L, Bracht Jorgensen H, Brady MV, Christensen S, de Ruiter PC, d'Hertefeldt T, Frouz J, Hedlund K, Hemerik L, Hol WHG, Hotes S, Mortimer SR, Setala H, Sgardelis SP, Uteseny K, van der Putten WH, Wolters V, Bardgett RD (2013) Soil food web properties explain ecosystem services across European land use systems. Proc Natl Acad Sci 1–6. 10.1073/pnas.1305198110

Foley JA, Ramankutty N, Brauman KA, Cassidy ES, Gerber JS, Johnston M, Mueller ND, O'Connell C, Ray DK, West PC, Balzer C, Bennett EM, Carpenter SR, Hill J, Monfreda C, Polasky S, Rockstrom J, Sheehan J, Siebert S, Tilman D, Zaks DPM (2011) Solutions for a cultivated planet. Nature 478(7369):337–342. http://www.nature.com/nature/journal/v478/n7369/abs/nature10452.html-supplementary-information

Fornara DA, Tilman D (2008) Plant functional composition influences rates of soil carbon and nitrogen accumulation. J Ecol 96(2):314–322. doi:10.1111/j.1365-2745.2007.01345.x

Godfray HCJ, Beddington JR, Crute IR, Haddad L, Lawrence D, Muir JF, Pretty J, Robinson S, Thomas SM, Toulmin C (2010) Food security: the challenge of feeding 9 billion people. Science 327(5967):812–818. doi:10.1126/science.1185383

Gomiero T, Pimentel D, Paoletti MG (2011) Environmental impact of different agricultural management practices: conventional vs. organic agriculture. Crit Rev Plant Sci 30:95–124. doi:10.1080/07352689.2011.554355

Haddad NM, Crutsinger GM, Gross K, Haarstad J, Tilman D (2011) Plant diversity and the stability of foodwebs. Ecol Lett 14(1):42–46. doi:10.1111/j.1461-0248.2010.01548.x

Holland JM (2004) The environmental consequences of adopting conservation tillage in Europe: reviewing the evidence. Agr Ecosyst Environ 103:1–25. doi:10.1016/j.agee.2003.12.018

Hooper DU, Chapin áV FS, Ewel JJ, Hector A, Inchausti P, Lavorel S, Lawton JH, Lodge DM, Lodge M, Loreau M, Naeem S, Schmid B, Setala H, Symstad AJ, Vandermeer J, Wardle DA (2005) Effects of biodiversity on ecosystem functioning: a consensus of current knowledge. Ecol Monogr 75(1):3–35

Hooper DU, Adair EC, Cardinale BJ, Byrnes JEK, Hungate BA, Matulich KL, Gonzalez A, Duffy JE, Gamfeldt L, O'Connor MI (2012) A global synthesis reveals biodiversity loss as a major driver of ecosystem change. Nature 1–5. doi:10.1038/nature11118

International Assessment of Agricultural Knowledge, Science and Technology for Development (IAASTD) (2009) Executive summary of the synthesis report. Island Press, Washington, 23 pp

Jones DL, Cross P, Withers PJA, DeLuca TH, Robinson DA, Quilliam RS, Harris IM, Chadwick DR, Edwards-Jones G (2013) Nutrient stripping: the global disparity between food security and soil nutrient stocks. J Appl Ecol 50(4):851–862. doi:10.1111/1365-2664.12089

Kiaer LP, Weisbach AN, Weiner J, Gibson D (2013) Root and shoot competition: a meta-analysis. J Ecol 101(5):1298–1312. doi:10.1111/1365-2745.12129

Kong AYY, Six J (2012) Microbial community assimilation of cover crop rhizodeposition within soil microenvironments in alternative and conventional cropping systems. Plant Soil 356(1–2): 315–330. doi:10.1007/s11104-011-1120-4

Lal R (2004) Soil carbon sequestration impacts on global climate change and food security. Science 304(5677):1623–1627. doi:10.1126/science.1097396

Lal R (2009) Soils and food sufficiency. A review. Agron Sustain Dev 29(1):113–133. doi:10.1051/agro:2008044

Miura T, Niswati A, Swibawa IG, Haryani S, Gunito H, Kaneko N (2013) No tillage and bagasse mulching alter fungal biomass and community structure during decomposition of sugarcane leaf litter in Lampung Province, Sumatra, Indonesia. Soil Biol Biochem 58:27–35

Montgomery DR (2004) Soil erosion and agricultural sustainability. Proc Natl Acad Sci 104(33):13268–13272. doi:10.1073/pnas.0611508104

Pérès G, Cluzeau D, Menasseri S, Soussana JF, Bessler H, Engels C, Habekost M, Gleixner G, Weigelt A, Weisser WW, Scheu S, Eisenhauer N (2013) Mechanisms linking plant community properties to soil aggregate stability in an experimental grassland plant diversity gradient. Plant Soil. doi:10.1007/s11104-013-1791-0

Robertson GP, Swinton SM (2005) Reconciling agricultural productivity and environmental integrity: a grand challenge for agriculture. Front Ecol Environ 3(1):38–46. doi:10.1890/1540-9295(2005)003[0038:rapaei]2.0.co;2

Seufert V, Ramankutty N, Foley JA (2012) Comparing the yields of organic and conventional agriculture. Nature 485:229–232. doi:10.1038/nature11069

Swaminathan MS (2006) An evergreen revolution. Crop Sci 46(5):2293–2303. doi:10.2135/cropsci2006.9999

Tilman D, Reich PB, Knops J, Wedin D, Mielke T, Lehman C (2001) Diversity and productivity in a long-term grassland experiment. Science 294(5543):843–845. doi:10.1126/science.1060391

Tilman D, Hill J, Lehman C (2006a) Carbon-negative biofuels from low-input high-diversity grassland biomass. Science 314(5805):1598–1600. doi:10.1126/science.1133306

Tilman D, Reich PB, Knops JMH (2006b) Biodiversity and ecosystem stability in a decade-long grassland experiment. Nature 441(7093):629–632. doi:10.1038/nature04742

Tilman D, Balzer C, Hill J, Befort BL (2011) Global food demand and the sustainable intensification of agriculture. Proc Natl Acad Sci U S A 108(50):20260–20264. doi:10.1073/pnas.1116437108

# Chapter 3
# Conservation and Sustainable Management of Soil Biodiversity for Agricultural Productivity

**Peter Wachira, John Kimenju, Sheila Okoth, and Jane Kiarie**

**Abstract** Soil biodiversity represents the variety of life belowground whose interaction with plants and small animals forms a web of biological activity. It improves the entry and storage of water, resistance to soil erosion, and plant nutrition, while also controlling soil pests and disease, and facilitating recycling of organic matter in the soil. Soil biodiversity is therefore the driver of healthy soil for sustainable crop production.

However, intensive agricultural activities are reported to lead to loss of soil biodiversity. This has been attributed to environmental degradation, and consequently to climate change. This paper highlights the importance of soil biodiversity and some factors associated with its loss, and presents a case study on selected soil organisms in Kenya. Results from this study indicated that land use changes affect soil biodiversity, and soil biodiversity determines the distribution of the aboveground biodiversity.

**Keywords** Biological control • Crop productivity • Soil biodiversity • Soil health • Sustainable utilization of soil

## 3.1 Introduction

Loss of biodiversity has been a major global topic over the past decade with the main reference being the big animals, plants, birds, and other visible organisms that are mainly above ground with little or no reference to soil biodiversity. Yet the highest percentage of life, far superseding the aboveground, is in the soil (Bardgett 2005). The soil is home to a large proportion of the world's genetic diversity of organisms,

P. Wachira (✉) • J. Kimenju • S. Okoth • J. Kiarie
School of Biological Sciences, University of Nairobi, Nairobi, Kenya
e-mail: pwachira@uonbi.ac.ke

N. Kaneko et al. (eds.), *Sustainable Living with Environmental Risks*,
DOI 10.1007/978-4-431-54804-1_3, 

amongst them microbes (fungi, bacteria, protozoa), mesofauna (nematodes, Acari, and Collembola), and macrofauna (arthropods, earthworms, and ants, among others), with the exception of megafauna (moles and rodents). It has been reported that in one gram of soil, there are over ten million microbes (Nannipieri et al. 1990). The food web is therefore incomplete without soil organisms. It is in fact fair to say that belowground biodiversity has not been given the prominence it deserves, despite the soil being the home of billions. Belowground biodiversity contributes greatly in the processes that form and stabilize the soil structure, decompose organic matter, control pests and disease, enable nutrient uptake by plants, and degrade harmful compounds in the soil. Macrofauna such as ants, termites, and earthworms are soil engineers, modifying soil structure to improve infiltration, retention, and availability of water and nutrients (Karanja et al. 2009). The process of conversion of organic matter to mineral elements is partly mediated by macrofauna. Soil microorganisms maintain soil functions such as decomposition, bioregulation of populations of plants and soil organisms, bioavailability of nutrients, detoxification, and bioremediation (Okoth 2004).

Current trends in climate change and the changing environment due to anthropogenic activities have tested the sustainability of the biosphere globally. Conversion of natural habitats for agriculture, mining, or construction is a major contributor to environmental degradation, an important precursor to climate change. These activities are carried out in an effort to provide food, shelter, and energy for the ever-increasing human population. Specifically, agricultural intensification has concentrated on gearing all effort toward maximum-yield production, including application of chemical fertilizers, pesticides, and extensive use of machinery, without any consideration for soil biodiversity. These practices have been identified as unsustainable and incompatible with nature, and their consequence is loss of soil biodiversity.

Loss of belowground biodiversity has been linked to an increase in soil pathogens, especially in agricultural ecosystems, leading to increased cost of production. Lack of awareness, knowledge, and understanding of belowground biodiversity has been identified as the major constraint on its management. Land intensification has challenged the conservation of soil biodiversity since it focuses only on increased crop yield with little or no regard to soil diversity. This leads to soil degradation and consequently to loss of soil biodiversity function. It is therefore important to conserve and sustainably utilize belowground biodiversity for increased agricultural productivity.

## 3.2 Soil Biodiversity and Its Importance to Agriculture

Biodiversity is usually defined as the variety and variability of living organisms and the ecosystems in which they occur. The soil is inhabited by a wide range of microorganisms (Davet and Francis 2000). Soil contains one of the most diverse assemblages of living organisms, although they are not visible to the naked eye (Giller et al. 1997). A typical healthy soil contains several species of vertebrate animals, several species of earthworms, 20–30 species of mites, 50–100 species of insects, tens of species of nematodes, hundreds of species of fungi, and perhaps thousands of species of bacteria and actinomycetes. Nannipieri et al. (1990) reported that in one gram of productive soil there can be over 100 million microorganisms. Hagvar (1998) noted that it is only in soil that organisms are densely packed in nature. Hawksworth and Mound (1991) reported some of the available estimates on the number of presently described species among selected soil biota that have been better studied. The number of soil-dwelling fungal species described ranges from 18 to 35,000, while the projected number may be greater than 100,000 (Hawksworth and Mound 1991). Nematodes and mites presently described comprise only 3 % and 5 %, respectively, of the total species (Hawksworth and Mound 1991). The estimates for bacteria and Achaea species are particularly problematic because of the differences in classification, and the present inability to culture many of these organisms (Hawksworth and Kalin-Arroyo 1995). However, it is a fact that these estimates are still preliminary and much lower than the estimated total number of species within each group. Soil biodiversity therefore reflects the mixture of living organisms in the soil. All these living things interact with one another and also with plants, forming a web of biological activity.

Soil microorganisms are very important in agriculture. Every chemical transformation taking place in the soil involves active contributions from each of them. In particular, they play an active role in soil fertility as a result of their involvement in the cycle of nutrients like carbon and nitrogen, which are required for plant growth (Muya et al. 2009). They are responsible for the decomposition of the organic matter entering the soil and therefore for the recycling of nutrients in soil (Okoth 2004). Other beneficial effects from soil microorganisms include: (1) organic matter decomposition and soil aggregation, (2) breakdown of toxic compounds including both metabolic by-products of organisms and agrochemicals, (3) inorganic transformations that make available nitrate, sulphate, and phosphate, as well as essential elements such as iron and manganese, and (4) nitrogen fixation into forms usable by higher plants (Anderson 1994). They have also been associated with management of soil pests and diseases. In summary, soil microorganisms improve the entry and storage of water, resistance to erosion, plant nutrition, and breakdown of organic matter. Other microorganisms will provide checks and balances to the food web through population control, mobility, and survival from season to season. In this regard, soil health has been defined as the capacity of a soil to function within ecosystem boundaries to sustain biological productivity, maintain environmental quality, and promote plant and animal health (Doran et al. 1996). Despite the important role played by soil biodiversity in ecosystems, anthropogenic processes are still providing challenges to the utilization and management of soil biodiversity.

## 3.3 Loss of Soil Biodiversity

The processes of land conversion and agricultural intensification are a significant cause of soil biodiversity loss. The factors controlling land conversion and agricultural intensification, and hence loss of soil biodiversity, are: population increase, national food insufficiency, internal food production imbalances, progressive urbanization, and a growing shortage of land suitable for conversion to agriculture. This has a negative consequence both on the environment and the sustainability of agricultural production. As land conversion and agricultural intensification occur, the planned biodiversity above ground is reduced (via monocultures) with the intention of increasing the economic efficiency of the system. This impacts the associated biodiversity of the ecosystem including microorganisms and invertebrate animals both above and below ground, thus lowering the biological capacity of the ecosystem for self-regulation, leading to further need for substitution of biological functions with agrochemicals and petrol energy inputs. The sustainability of these systems thus comes to depend on external and market-related factors rather than internal biological resources. Other practices that lead to loss of biodiversity are continuous cultivation of land without a period of rest, monoculture, removal of crop residues by burning or transfer for use as fodder, soil erosion, soil compaction due to degradation of the soil structure, and repeated application of pesticides.

Changes in the belowground biodiversity are often thought to reflect those above ground (Wall and Nielsen 2012). There is evidence that the soil community may be more functionally resilient than the aboveground biota. It is often hypothesized that reduction in the diversity of the soil community, including cases of species extinction, may cause a catastrophic loss in function, reducing the ability of an ecosystem to retain its self-perpetuating characteristic.

## 3.4 Management and Conservation of Soil Biodiversity: A Case Study in Kenya

Management strategies encompassing minimum tillage, crop rotation, and incorporation of crop residues and manure, alter the soil quality and the capacity of soil to perform its functions. Farming practices that minimize soil disturbance (plowing) and return plant residues to the soil allow gradual restoration of soil organic matter. Reduced tillage also tends to increase build-up of beneficial organisms. On the other hand, soil compaction, poor vegetation cover and/or lack of plant litter covering the soil surface tend to reduce the number of beneficial soil organisms.

Based on previous research and experiences with regard to management of belowground biodiversity in Kenya, a survey of selected belowground organisms was monitored over a land use gradient. The study was based on the hypothesis that beneficial soil microorganisms decline with land use intensification (Vandermeer et al. 1998). The land use types ranged from natural forest to horticulture farms through maize bean systems and fallow. The natural forest represented the undisturbed systems while the horticulture farms represented the intensively cultivated land use types. The magnitude of land use intensification was determined by tillage, application of fertilizers, pesticides, and herbicides, as well as regular turning of soil. The forests were therefore the least disturbed land uses and were regarded as the benchmark while the horticulture was regarded as the most intensively cultivated

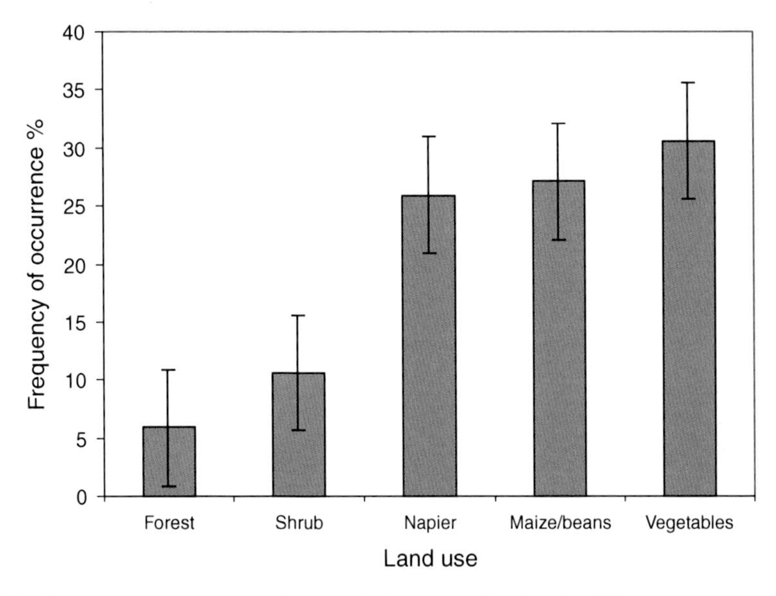


**Fig 3.1** Frequency of occurrence of nematode-destroying fungi under different land use types in Taita Taveta, Kenya (Wachira et al. 2009)

**Table 3.1** Comparison of plant parasitic nematodes (PPN), free-living nematodes (FLN), and ratio of free-living to plant parasitic nematodes in soils under different uses in Taita Taveta, Kenya

| Land use | PPN | FLN | FLN:PPN |
|---|---|---|---|
| Natural forest | 189 | 759 | 4.01 |
| Planted forest | 353 | 526 | 1.49 |
| Fallow | 796 | 458 | 0.57 |
| Coffee | 948 | 419 | 0.44 |
| Maize | 938 | 120 | 0.13 |
| Vegetable | 915 | 191 | 0.21 |
| LSD ($p<0.05$) | 122 | 137 | |

land use. In this study, soil microorganisms (represented by selected fungal groups), mesofauna (represented by nematodes, Acari, and Collembola) and macrofauna (represented by ants and earthworms) were studied. Laboratory and field extractions were conducted for the microorganisms and macrofauna respectively.

A general build-up of soil pathogens (*Pythium, Fusarium,* and *Rhizoctonia*) was observed as land intensification increased. The study also revealed that land use intensification (land disturbance) negatively influences the abundance and species richness of soil Collembolan communities. In particular, the nematode population and diversity were higher in natural forest than in the horticulture farms. A decline in beneficial microorganisms, for example, entomopathogenic nematodes, was also observed as land intensification increased, except for nematode-destroying fungi whose diversity increased with land use intensification (Fig. 3.1). The ratio of plant parasitic nematodes to free living nematodes under vegetable land use (0.21) was lower than the ratio recorded in natural forest, which was 4.01 (Table 3.1). Comparing land uses under crops, those with perennial crops like coffee recorded

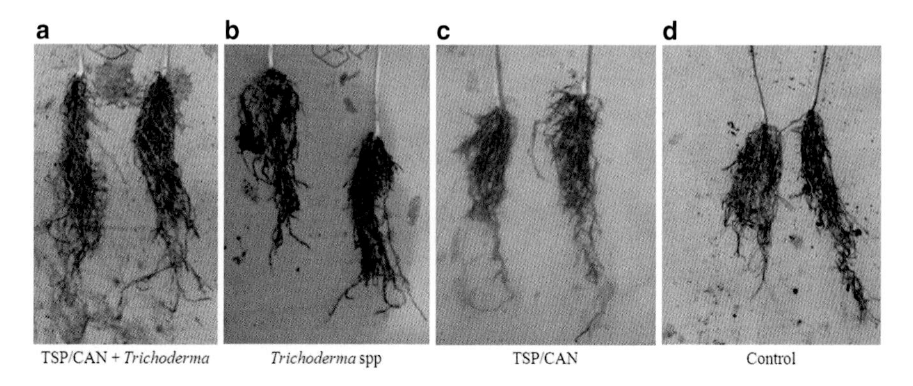


**Fig 3.2** Effect of *Trichoderma* on root growth of bean seedlings. (**a**) TSP/CAN+*Trichoderma*, (**b**) *Trichoderma* spp, (**c**) TSP/CAN, (**d**) Control

high population and diversity of soil microorganisms compared to land uses under annual crops. The results from this study generally revealed that the undisturbed land uses had higher soil biodiversity both in abundance and diversity compared to the disturbed.

Best-bet practices were selected and tested under field conditions on their ability to retain soil biodiversity and improve the crops. They included bio-inoculants (*Trichoderma*, Bacillus, Arbuscular mycorrhiza fungi (AMF), fertilizers (Mavuno and triple super phosphate (TSP) + calcium ammonium nitrate (CAN)), cow manure, and combinations involving inoculants and fertilizers or manure. The potential of soil fertility management practices such as Mijingu + CAN, Mavuno, manure, and TSP + CAN were also demonstrated. The efficacy of bio-inoculants, especially when combined with manure and fertilizers, had a positive and synergistic effect on yield in maize. Compared to other treatments in this study, application of TSP + CAN combined with *Trichoderma* enhanced root growth in beans leading to better anchorage and nutrient supply (Fig. 3.2).

From this research, it was reported that increased soil biodiversity results in increased capacity of the soil to control soilborne pathogens and pests. This reflects the fact that some of the organisms, which respond to addition of organic substrates, are antagonistic to the pathogenic ones. In particular, application of manure led to an increase in the soil of nematode-trapping fungi, natural enemies of plant parasitic nematodes that capture and kill nematodes (Wachira et al. 2009). As a result of this study, it is recommended that repeated application of the bio-inoculants should be encouraged to build up their numbers in the soil (Okoth et al. 2011). Another management strategy is the use of crop rotation. Crop rotation is a widely known practice that helps to break the life cycles of plant pathogens and pests. Unrelated crops are usually selected and grown in succession. Planting several crops (intercropping as opposed to monoculture) at the same time or in relay cropping systems helps to reduce build-up of one group of organisms at the expense of others. Examples of intercrop systems that have value in building up soil biodiversity and controlling harmful organisms include maize/bean, cabbage/cowpea, spider plant/egg plant.

## 3.5 Conclusion

Soil organisms are the primary agents of nutrient cycling, and hence of food and fiber production; they therefore determine the occurrence and distribution of the aboveground biodiversity. The soil fauna plays a major role in modification of the soil structure that in turn regulates water dynamics. Despite their functional role, soil biota remain an unexplored area, and scientific understanding is therefore lacking. One of the reasons is inadequate knowledge and understanding regarding the appropriate methods to study these microorganisms due to their enormous numbers. Therefore agricultural activities that promote their population in the soil should be promoted for crop production.



## References

Anderson J (1994) Function attributes of biodiversity in land use systems. In: Greenland DJ, Szabolcs I (eds) Soil resilience and sustainable land use. CAB International, Wallingford, pp 267–290

Bardgett RD (2005) The biology of soil: a community and ecosystem approach. Oxford University Press Inc, New York

Davet P, Francis R (2000) Detection and isolation of soil fungi. Science Publishers, Enfield New Hampshire

Doran JW, Sarrantonio M, Liebig MA (1996) Soil health sustainability. In: Sparks DL (ed) Advances in agronomy, vol 56. Academic, San Diego, pp 1–54

Giller K, Beare MH, Izac AM, Swift MJ (1997) Agricultural intensification, soil biodiversity and agroecosystem function. Appl Soil Ecol 6:3–16

Hagvar S (1998) The relevance of the Rio-Convention on biodiversity to conserving the biodiversity of soils. Appl Soil Ecol 9:1–7

Hawksworth DL, Kalin-Arroyo MT (1995) Magnitude and distribution of biodiversity. In: Heywood VH (ed) Global biodiversity assessment. Cambridge University Press, Cambridge, pp 107–191

Hawksworth DL, Mound LA (1991) Biodiversity databases: the crucial significance of collections. In: Hawksworth DL (ed) The biodiversity of microorganisms and invertebrates: its role in sustainable agriculture. CAB International, Wallingford, pp 17–29, 302 pp

Karanja NK, Ayuke FO, Muya EM, Musombi BK, Nyamasyo GHN (2009) Soil macrofauna community structure across land use systems of Taita, Kenya. Trop Subtrop Agroecosyst 11:385–396

Muya E, Karanja N, Okoth PZ, Roimen H, Mung'atu J, Motsotso B et al (2009) Comparative description of land use and characteristics of belowground biodiversity benchmark sites in Kenya. Trop Subtrop Agroecosyst 11:263–275

Nannipieri P, Grego S, Ceccanti B (1990) Ecological significance of the biological activity in soil. Soil Biochem 6:293–355

Okoth SA (2004) An overview of the diversity of microorganisms involved in the decomposition in soils. J Trop Microbiol 3:3–13

Okoth SA, Jane AO, James OO (2011) Improved seedling emergence and growth of maize and beans by Trichoderma harzianum. Trop Subtrop Agroecosyst 13:65–71

Vandermeer J, Noordwijk M, Anderson JM, Ong C, Perfecto I (1998) Global change and multi-species agro ecosystems: concepts and issues. Agric Ecosyst Environ 67:1–22

Wachira PM, Kimenju JW, Okoth SA, Mibey RK (2009) Stimulation of nematode-destroying fungi by organic amendments applied in management of plant parasitic nematode. Asian J Plant Sci 8:153–159

Wall DH, Nielsen UN (2012) Biodiversity and ecosystem services: is it the same below ground? Nat Educ Knowl 3(12):8

# Chapter 4
# Conservation Tillage Assessment for Mitigating Greenhouse Gas Emission in Rainfed Agro-Ecosystems

**Muhajir Utomo**

**Abstract** Global warming due to greenhouse gas emissions is currently receiving considerable attention worldwide. Agricultural systems contribute up to 20 % of this global warming. However, agriculture can reduce its own emissions while increasing carbon sequestration through use of recommended management practices, such as consernvation tillage (CT). The objective of this paper is to review the role of long-term CT in mitigating greenhouse gas emissions during corn production in rainfed tropical agro-ecosystems. The types of conservation tillage were no-tillage (NT) and minimum tillage (MT). In a long-term plot study, $CO_2$ emission from CT throughout the corn season was consistently lower than that from intensive tillage (IT). The cumulative $CO_2$ emissions of NT, MT, and IT in corn crops were 1.0, 1.5, and 2.0 Mg $CO_2$-C $ha^{-1}season^{-1}$, respectively. Soil carbon storage at 0–20 cm depth after 23 years of NT cropping was 36.4 Mg C $ha^{-1}$, or 43 % and 20 % higher than the soil carbon strorage of IT and MT, respectively. Thus, NT had sequestered some 4.4 Mg C $ha^{-1}$of carbon amounting to carbon sequestration rate of 0.2 Mg C $ha^{-1}$ $year^{-1}$. IT, on the other hand, had depleted soil carbon by as much as 6.6 Mg C $ha^{-1}$, yielding a carbon depletion rate of 0.3 Mg C $ha^{-1}$ $year^{-1}$. Assessment of the farmer's corn fields confirmed these findings. $CO_2$ emission from CT corn farming was similar to that of rubber agroforest and lower than IT corn farming. Based on carbon balance analysis, it can be concluded that corn crops in tropical rainfed agro-ecosystems were not in fact net emitters, and that NT was a better net sinker than other tillage methods.

**Keywords** Conservation tillage • $CO_2$ emission • Carbon storage

M. Utomo (✉)
Agro-Technology Department, Faculty of Agriculture, University of Lampung,
Jl. Sumantri Brojonegoro #1, Bandar Lampung 35145, Indonesia
e-mail: mutomo2011@gmail.com

N. Kaneko et al. (eds.), *Sustainable Living with Environmental Risks*,
DOI 10.1007/978-4-431-54804-1_4, 

## 4.1 Introduction

Global warming due to greenhouse gas (GHG) emissions is currently receiving consider-able attention worldwide. The impact of human activities on the atmosphere and the accompanying risk of long-term climate change on a global-scale are by now familiar topics to many people (Paustian et al. 2006). Global temperature rose 0.6 °C during the twentieth century, and is projected to increase by 1.5–5.8 °C during the twenty-first century. Historical records clearly show an accelerating increase in atmospheric GHG concentrations over the past 150 years (Intergovernmental Panel on Climate Change (IPCC) 2001). This is attributed to the advance of greenhouse gases such as $CO_2$, $CH_4$, and $N_2O$, in particular, due to the anthropogenic activities. Among the greenhouse gases, $CO_2$ is the most important gas, accounting for 60 % of global warming (Rastogi et al. 2002; Ruddiman 2003; Lal 2007). While most of the increase is due to $CO_2$ emissions from fossil fuels, land use and agriculture play significant roles. Overall, agricultural activities along with land use change, which predominantly occurs in the tropics, globally account for about one-third of the warming effect from increased GHG concentrations (Cole et al. 1997). In fact, although agriculture is its self subject to environmental risk due to global warming, ironically it is also estimated to contribute up to 20 % of global anthropogenic $CO_2$ emissions (Intergovernmental Panel on Climate Change (IPCC) 2006; Haile-Mariam et al. 2008). In Indonesia specifically, agriculture, land use change, and forests combine to contribute as much as 53 % of $CO_2$ emissions (Boer 2010). Agro-ecosystems emit $CO_2$ emission through direct use of fossil fuels in food production, indirect use of embodied energy in inputs, and cultivation of soils that cause the loss of carbon through decomposition and erosion (Ball and Pretty 2002).

The difference compared with fossil fuel based sectors, however, is that land use and agriculture have the opportunity to mitigate GHG emission through recommended management practices (RMP). Therefore, producers, scientists, and planners are faced with the challenge of increasing agricultural production without aggravating the risks of GHG emissions. In this regard, the management of soil resources in general and that of soil organic carbon (SOC) in particular, is extremely important. The world's soil resources may be the key factor in the creation of an effective carbon sink and mitigation of the greenhouse effect (Lal 1997). By employing RMP, agro-ecosystems can act as sinks that can both sequester carbon (C) and reduce $CO_2$ emission (Pretty and Ball 2001; Lal 2007). Conservation tillage as a RMP can enhance SOC, thus reducing agriculture's potential for global warming (Rastogi et al. 2002; Lal 2007; Smith 2010). In fact, in the Kyoto Climate Protocol and IPCC Guidelines for National Greenhouse Gas Inventories, conservation tillage is listed as an option for carbon sequestration (Sedjo et al. 1998; Eggleston et al. 2006).

Worldwide adoption of CT, and particularly no-tillage, has expanded rapidly since about 1990, particularly in the United States, South American countries, and Africa (Triplett and Dick 2008). As in other countries, CT in Indonesia which generally consists of no-tillage (NT) and minimum tillage (MT), was initially promoted by a few CT researchers in the 1980's. Farmers themselves successfully adopted and practiced CT in the 1990s due to the fact that it requires less cost and labor, yet maintains at least the same crop yield as IT. This was the case particularly in regions with labor shortages, such as Sumatra, Borneo, and Celebes (Utomo 2004). Then in 1998, CT was explicitly advocated in a national land preparation policy, resulting in increasing adoption of the techniques, particularly for corn production (Utomo et al. 2010a). As the second most important food crop in Indonesia, corn is mostly planted in rainfed agro-ecosystems. In Lampung Province, the area of corn harvested in 2011 was 380.917 ha, or 46 % of the total area of Sumatra's corn belt (Badan Pusat Statistik BPS 2012). However, rainfed agro-ecosystems, which account for about 91 % of total agricultural land in Indonesia, are inherently prone to degradation. To sustain these vulnerable agro-ecosystems, therefore, CT should be implemented and further improved.

The aim of this paper is to review research and assessment findings both from a long-term plot and from farmers' fields, in order to evaluate the potential of CT to mitigate $CO_2$ emissions in Indonesia's rainfed agro-ecosystems. In this paper, mitigation of $CO_2$ emissions is defined as a technological effort both to reduce GHG emissions and to sequester carbon in soils.

## 4.2 Soil, Carbon Dioxide Emission, and Conservation Tillage

Soil is a powerful natural sink of carbon in terrestrial ecosystems. Natural soils can retain carbon in stable microaggregates for up to hunded and thousands of years unless environmental conditions are changed and stable soil structure is damaged. Cultivation practices, such as plowing, break soil aggregates, exposing formerly protected SOC in soil to microbial attacks, and thus accelerating decomposition and $CO_2$ emission to the atmosphere (Luo and Zhou 2006). In general, these respiratory carbon losses from soil can be attributed to biological and chemical processes within the soil that may include $CO_2$ from soil organic matter and crop residue decomposition, and from root respiration (Rastogi et al. 2002; Al-Kaisi and Yin 2005). Moreover, Luo and Zhou (2006) stated that $CO_2$ emitted from soil ecosystems constitutes part of the cabon cycle, and is mostly produced as a result of the soil respiration process. Depending on the sources of carbohydrate substrate supply, $CO_2$ production in the soil can be attributed to root respiration, microbial respiration in the rhyzosphere, litter decomposition, and oxidation of soil organic matter.

In tropical agro-ecosystems, soil respiration and decomposition happen more quickly, resulting in higher $CO_2$ emission and less C sequestration than in cooler climates (Desjardins, et al. 2002). Cultivation for land preparation produces a favorable soil microenvironment that can accelerate microbial decomposition of plant residues. Cultivation or intensive tillage (IT) is any tillage that requires clean and loose top soil for seed to grow. For this reason, soil should be totally tilled and no mulch is needed. But over the long-term, IT decreases soil quality and soil productivity (Rastogi et al. 2002; Paustian et al. 2006; Luo and Zhou 2006). Soil degraded by cultivation is also more susceptible to erosion, which carries carbon to rivers and oceans, where it is partially released into the atmosphere by outgassing (Luo and Zhou 2006).

Soil resources have the potential capacity to sequester carbon. Based on the principles of either increasing plant carbon input or slowing soil carbon decomposition rates, soil carbon can be sequestered through a variety of recommended management practices (RMP). Conservation tillage as a RMP is a tillage system that keeps at least 30 % of the soil surface covered by plant residue and reduces soil disturbance (Lal 1989; Utomo 2004). The function of crop residue covering the soil surface is to protect the soil from sun, rain, and wind, and to feed the biota. Crop residue serves as a substrate that is converted to microbial biomass and soil organic matter, and has the potential to enhance carbon sequestration in agricultural soils (Wright and Hons 2004). There are several types of CT, including (a) *no-tillage*: the soil is left undisturbed except for hills, slots, or bands; and weeds are controlled primarily with herbicide; (b) *ridge tillage*: soil is undisturbed, and planting is on ridges; (c) *strip tillage*: soil is undisturbed, and 1/3 of the soil surface is tilled; (d) *mulch* tillage: soil is totally tilled, with mulch on the soil surface; and (e) *reduced tillage/minimum tillage*: at least 30 % of the soil surface is covered by plant residue (Lal 1989; Utomo 2004).

Long-term CT involving crop residue and less tillage can reduce soil erosion and improve soil organic matter. Therefore, through its effect on C dynamics, aggregation, and soil structure, and its interaction with cropping systems, CT is expected to result in lower $CO_2$ emissions and higher soil C sequestration than IT (Lal 1997).

## 4.3 Reducing Carbon Dioxide Emission

### *4.3.1 Carbon Dioxide Emission at the Long-Term Plot*

Field research on mitigation of $CO_2$ gas emissions from a corn plot was conducted from 2009 to 2011 as part of the long-term plot research commenced in 1987 in Lampung, Indonesia (105° 13′E, 05° 21′S). The experiment was a factorial, randomized complete block design, with 4 replications. Tillage treatments comprised conservation tillage (NT and MT), and IT, while nitrogen fertilization rates were 0, 100, and 200 kg N ha$^{-1}$ (Utomo et al. 1989).

Regardless of N fertilization, average $CO_2$-C emission from tillage treatment measured before plowing was 3.3 kg $CO_2$-C ha$^{-1}$day$^{-1}$. It appears that just one day after plowing (1 DAP), $CO_2$-C emission from IT increased sharply to reach a maximum magnitude of 14.6 kg $CO_2$-C ha$^{-1}$ day$^{-1}$. Thereafter, $CO_2$-C emission from IT dropped sharply at 3 DAP and then gradually declined, while emission from CT was relatively level to the end of the season (Fig. 4.1a) (Utomo et al. 2012). This was similar to research findings by Al-Kaisi and Yin (2005), which found that $CO_2$ emission was generally lower with less tillage compared to moldboard plow usage, with the greatest differences occurring immediately after tillage operations.

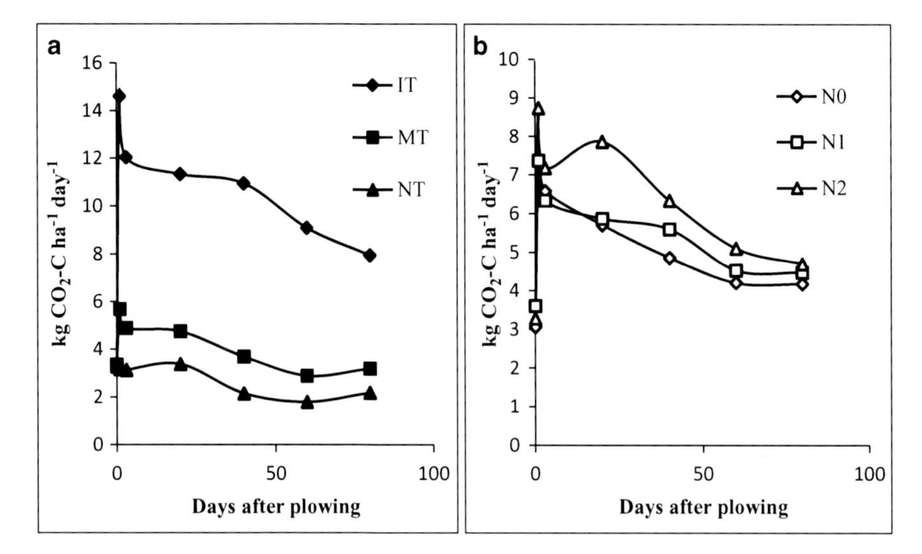


**Fig. 4.1** Pattern of $CO_2$-C emission in corn season as affected (**a**) conservation tillage, and (**b**) N fertilization; IT = intensive tillage, MT = minimum tillage, NT = no-tillage, N0 = 0 kg N ha$^{-1}$, N1 = 100 kg N ha$^{-1}$, N2 = 200 kg N ha$^{-1}$ (Utomo et al. 2012)

During a single season, NT and MT reduced the $CO_2$-C emissions of corn production at the long-term plot to 76 % and 62 % of IT based emission, respectively. This was because tillage broke and inverted the soil to allow rapid $CO_2$ loss and $O_2$ entry, and mixed together the residues and organic particles that could enhance microbial attack (Reicosky 2001; Rastogi et al. 2002; Smith and Collins 2007). On the other hand, CT reduced gas diffusivity and air-filled porosity, and kept SOC unexposed, resulting in a lower $CO_2$ emission than that of IT (Rastogi et al. 2002). These findings are in agreement with those reported by Reicosky (2001); Desjardins et al. (2002); Scala et al. (2005); Brye et al. (2006).

Although the effects were not as strong as those of tillage treatment, N fertilization treatment in corn season also consistently increased $CO_2$-C emission (Fig. 4.1b). Emissions of $CO_2$ at the 200 kg N ha$^{-1}$ fertilization rate were consistently higher than those at the 0 and 100 kg N ha$^{-1}$ rates (Utomo et al. 2012). When tillage was combined with N fertilization, the synergetic effect was clearly observed. With residual 200 kg N ha$^{-1}$, $CO_2$ emission from IT treatment at 1 DAP was the highest among treatment combinations, while MT with any N rate fertilizations produced the second highest $CO_2$ emission, and NT was the lowest.

The higher $CO_2$-C emission when combining IT with a higher N rate was associated with the synergetic effect of tillage and N fertilization treatments. Combination of IT and an optimum N rate created a soil micro climate and available N that produced more soil $CO_2$ emission (Utomo et al. 2012).

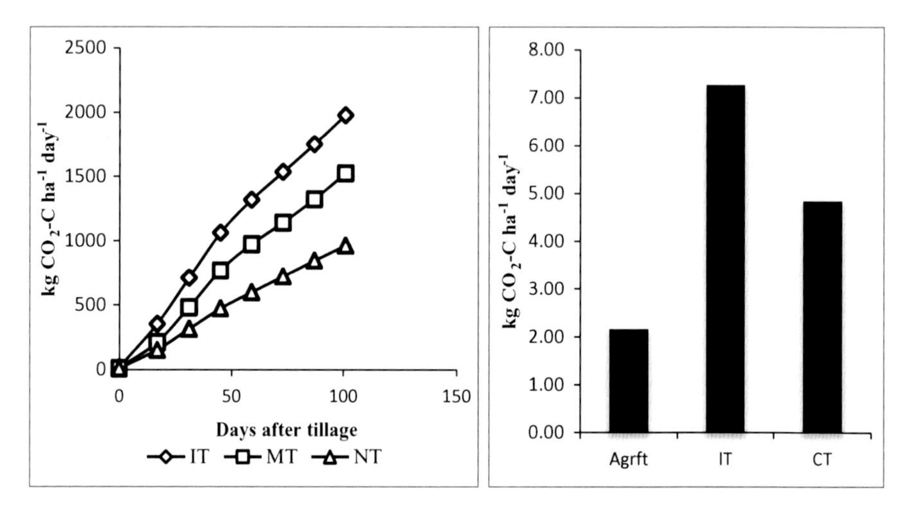


**Fig. 4.2** Cumulative $CO_2$-C emission of corn at long-term plot (*left*) and $CO_2$-C emission at farmers' fields (*right*); Agrft = agroforest, IT = intensive tillage, MT = minimum tillage, NT = no-tillage and CT = conservation tillage (Utomo et al. 2010b; Utomo et al. 2011)

## 4.3.2 Cumulative $CO_2$ Emission at the Long-Term Plot

Cumulative soil $CO_2$ emission was set using the equation proposed by Al-Kaisi and Yin (2005). Cumulative soil $CO_2$ emissions of IT, MT, and NT were 1.98, 1.53 and 0.96 Mg $CO_2$-C $ha^{-1}$ $season^{-1}$, respectively (Fig. 4.2, left). During a single season, NT reduced $CO_2$ emission to 52 % of IT based emission, while MT reduced emission to 23 % that of IT (Utomo et al. 2011).

Although these figures are somewhat lower than those of the average bases method, the value of cumulative $CO_2$ emission is much closer to continuous $CO_2$ measurement. This finding is in accordance with findings reported by Al-Kaisi and Yin (2005). They reported that cumulative soil $CO_2$ emission from MT was 19 to 41 % lower than that from moldboard plow usage, and NT with residue was 24 % lower than NT without residue during the 480-h measurement period.

## 4.3.3 Carbon Dioxide Emission Assessment in Farmers' Fields

In 2010, assessment of $CO_2$ emission in farmers' fields was conducted in East Lampung District, Lampung Province, Indonesia (105°28′35″–105°28′39″E, 05°19′22″–05°19′26″S). The soil texture was loam to clay loam, with soil pH $_{H2O}$ 5.1–5.4, total soil N 0.15–19 %, soil organic C 0.7–1.0 %, available P 1.9–4.1 ppm, CEC 10.2–13.2 me 100 $g^{-1}$, and BD 1.2–1.3 Mg $m^{-3}$ (Utomo et al. 2010b).

In this assessment, a similar effect was clearly shown, but the effect was not as marked as in the plot experiment (Fig. 4.2, right). This was not only because the farmer applied less fertilizer, but also because during MT farming mulch covered only around 40 % of the soil surface, while in the plot experiment it covered around 90 %. Emission of $CO_2$ from IT was the highest, while emission from rubber agroforest was the lowest (Utomo et al. 2010b). Rubber agroforest reduced $CO_2$ emission to 70 % that of IT farming, while MT farming reduced it as much as 33 % (Fig. 4.2, right).

## 4.4 Enhancing Carbon Sequestration

### *4.4.1 Soil Carbon Storage*

At the long-term plot, the highest soil C storage after 23 years of cropping at 0–20 cm depth was obtained by treatment combining NT with a higher N rate, while the lowest soil C strorage was in IT with 0 kg N/ha as shown in Fig. 4.3, left. No-tillage and MT resulted in soil C storage 43 % and 20 % higher than IT, respectively. The initial carbon storage at 0–20 cm depth in 1987 (when this long-term plot was established) was 32.0 Mg $ha^{-1}$(Utomo et al. 2010a). Thus, during 23 years of cropping, NT had sequestered as much as 4.4 Mg C $ha^{-1}$of carbon, amounting to a carbon sequestration rate of 0.2 Mg C $ha^{-1}$ $year^{-1}$. In contrast, IT had depleted 6.6 Mg C $ha^{-1}$ of carbon, yielding with carbon depletion rate of 0.3 Mg C $ha^{-1}$ $year^{-1}$. The higher C sequestration of CT than business as usual practice was attributed to addition of previous plant residues, and a lower rate of soil organic matter decomposition with respect to CT. Every season, the average weight of crop residue applied to the NT soil surface was 6–13 Mg $ha^{-1}$ $season^{-1}$ with a C-N ratio of around 32 (Utomo et al. 2010a).

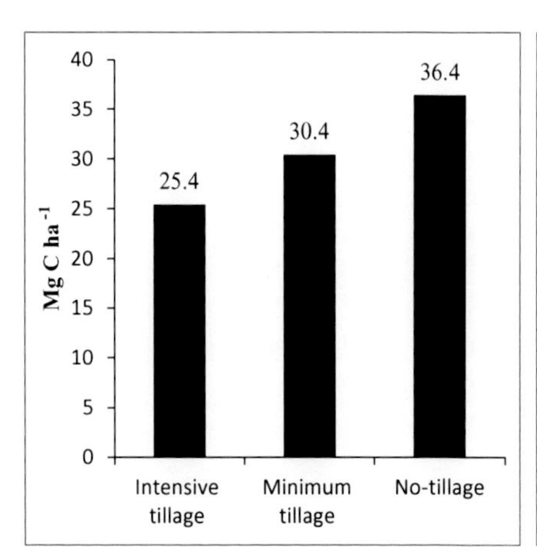


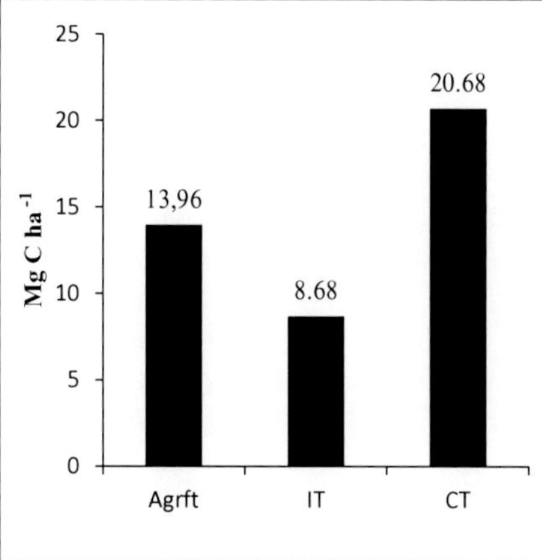


**Fig. 4.3** Soil carbon storage at 0–20 cm depth after 23 years of conservation tillage (*left*) and farmers' fields (*right*); Agrft = agroforest, IT = intensive tillage, MT = minimum tillage, NT = no-tillage and CT = conservation tillage ( Utomo et al. 2010b)

**Table 4.1** Carbon balance of corn (during a single season)

| Treatment | Root | Stalk | Grain | Total C-biomass (Mg C ha$^{-1}$) | Emission | Net sequestration |
|---|---|---|---|---|---|---|
| Intensive tillage | 1.2 | 3.6 | 2.5 | 7.3 | 2.0 | 5.3 |
| Minimum tillage | 1.6 | 3.4 | 4.5 | 9.5 | 1.5 | 8.0 |
| No-tillage | 2.1 | 5.3 | 5.0 | 12.4 | 1.0 | 11.4 |

*Note*: With optimum fertilization (Utomo et al. 2011)

This higher soil carbon sequestration is also reflected in improved soil quality and crop productivity with respect to CT. Utomo et al. (2013) recently reported that compared to the IT corn field, the CT corn field after 23 years of cropping had higher soil moisture, soil exchange bases, and soil microbial biomass. The corn yield of long-term CT was also 31.8 % higher than that of IT.

At the farmer's fields, that finding was confirmed by soil C storage at 0–20 cm depth under the different land use systems presented in Fig. 4.3, right. Soil C storage under CT farming was 138 % higher than under IT farming and 48 % higher than under rubber agroforest. The significant increase in soil C storage was attributable to the decomposition of previous crop residues and less soil erosion with respect to CT and rubber forest (Utomo et al. 2010b).

### 4.4.2 *Carbon Sequestration of Corn Crops*

Carbon sequestration of corn biomass was measured at harvest time. Through photosynthesis, plants fix $CO_2$ from the air and convert it into organic carbon compounds that are used to grow plant tissues or biomass (Luo and Zhou 2006). The total carbon of NT corn biomass was 12.4 Mg C ha$^{-1}$, 31 % higher than MT and 70 % higher than IT. With a better micro-climate and soil quality (Utomo et al. 2013), CT sequestered carbon in biomass at a higher level than other tillage systems, as reported by Lal (1997), Wright and Hons (2004), and Smith and Collins (2007). As shown in Table 4.1, NT's potential net sequestration reached 11.4 Mg C ha$^{-1}$, or 115 % and 43 % higher than IT and MT, respectively.

Dispite the fact that tillage systems generated $CO_2$ emissions, howevr, all tillage systems also seqestered carbon at a rate higher than their $CO_2$ emissions (Table 4.1). Thus, CT corn is not in fact a net $CO_2$ emitter, but instead is a net sinker. In the final analysis, therefore, it is evident that CT farming using RMP can mitigate $CO_2$ emission in a rain-fed tropical agro-ecosystem.

## 4.5 Conclusions and Policy Implication

In tropical rainfed agro-ecosystems, long-term conservation tillage of corn reduced $CO_2$ emission and increased carbon sequestration both in biomass and soil. Long-term conservation tillage of corn was also an effective net sinker of carbon.

However, further research is needed to improve the capacity of conservation tillage technology to mitigate greenhouse gas emissions in other crops and in different agro-ecosystems.

The policy implication of this strategic finding is that conservation tillage should be promoted by farmers, policy makers, and politicians as a recommended management practice for halting environmental degradation, reducing greenhouse gas emission, and strengthening food security.



## References

Al-Kaisi MM, Yin X (2005) Tillage and crop residue effects on soil carbon and carbon dioxide emission in corn–soybean rotations. J Environ Qual 34:437–445

Badan Pusat Statistik (BPS) (2012) Luas panen, produktivitas dan produksi tanaman jagung seluruh provinsi. Sensus Pertanian, Republik Indonesia 2013 (in Indonesian)

Ball AS, Pretty JN (2002) Agricultural influences on carbon emissions and sequestration. University of Essex, Colchester

Boer R (2010) Strategi mitigasi emisi GRK dari lahan. In: Centre for climate risk and opportunity management in South East Asia and Pacific (CCROM SEAP). Bogor Agricultural University, Kota Bogor

Brye KR, Longer DE, Gbur EE (2006) Impact of tillage and residue burning on carbon dioxide flux in a wheat–soybean production system. Soil Sci Soc Am J 70:1145–1154. SSSA, Madison, USA

Cole CV, Duxbury J, Freney J, Heinemeyer O, Minami K, Mosier A, Paustian K, Rosenberg N, Sampson N, Sauerbeck D, Zhao Q (1997) Global estimates of potential mitigation of greenhouse gas emissions by agriculture. Nutr Cycl Agroecosyst 49:221–228

Desjardins RL, Smith W, Grant B, Campbell C, Janzen H, Riznek R (2002) Management strategies to sequester carbon in agricultural soils and to mitigate greenhouse gas emissions. In: International workshop on reducing vulnerability of agriculture and forestry to climate variability and climate change. Ljubljana, 7–9 Oct 2002

Eggleston S, Buendia L, Miwa K, Ngara T, Tanabe K (2006) IPCC guidelines for national greenhouse gas inventories. Technical Report, IPCC

Haile-Mariam S, Collins HP, Higgins SS (2008) Greenhouse gas fluxes from an irrigated sweet corn (*Zea mays* L.)–potato (*Solanum tuberosum* L.) rotation. J Environ Qual 37:759–771

Intergovernmental Panel on Climate Change (IPCC) (2001) Climate change 2001: the scientific basis. Contribution of working group I to the third assessment report of the Intergovernmental Panel on Climate Change. Cambridge University Press, Cambridge

Intergovernmental Panel on Climate Change (IPCC) (2006) Guidelines for national greenhouse gas inventories. IPCC/IGES, Hayama

Lal R (1989) Conservation tillage for sustainable agriculture: tropics versus temperate environment. Adv Agron 42:85–197

Lal R (1997) Residue management, conservation tillage and soil restoration for mitigating greenhouse effect by $CO_2$-enrichment. Soil Tillage Res 43:81–107

Lal R (2007) Soil carbon sequestration to mitigate climate change and advance food security. Soil Sci 32(12):943–956

Luo Y, Zhou X (2006) Soil respiration and the environment. Academic/Elsevier, Burlington, p 316

Paustian K, Antle JM, Sheehan J, Paul EA (2006) Agriculture's role in greenhouse gas mitigation. Pew Center on Global Climate, Arlington, VA, USA, p 76

Pretty JN, Ball AS (2001) Agricultural influences on carbon emission and sequestration: a review of evidence and emerging trading options. In: Centre for environment and society occasional paper 2001–2003. University of Essex, Colchester, p 30

Rastogi M, Singh S, Pathak H (2002) Emission of carbon dioxide from soil. Curr Sci 82(5): 510–517

Reicosky DC (2001) Effects of conservation tillage on soil organic carbon dynamics: field experiment in the U.S. corn belt. In: Scott DE, Mohtar RH, Steinhart GC (eds) Sustaining the global farm. Purdue University and the USDA-ARS National Soil Erosion Research Laboratory, Morris, MN, USA, pp 481–485

Ruddiman WF (2003) The anthropogenic greenhouse era began thousands of years ago. Clim Change 61:261–293

Scala L, Bolonhezi ND, Pereira GT (2005) Short-term soil $CO_2$-C emission after conventional and reduced tillage of a no-till sugar cane area in southern Brazil. Soil Tillage Res 91(1–2): 244–248

Sedjo R, Sohngen B, Jagger P (1998) Carbon sinks in the post-Kyoto world, Internetth edn. Resources for the Future, Washington DC

Smith KE (2010) Effect of elevated $CO_2$ and agricultural management on flux of greenhouse gases from soil. Soil Sci 175(7):349–356

Smith JL, Collins HP (2007) Management of organisms and their processes in soils. In: Paul EA (ed) Soil microbiology, ecology and biochemistry, 3rd edn. Academic, Burlington, p 532

Triplett GB, Dick WA (2008) No-tillage crop production: a revolution in agriculture. Agro J 100:S153–S156

Utomo M (2004) Olah tanah konservasi untuk budidaya jagung berkelanjutan. In: Prosiding Semi-nar Nasional IX Budidaya Pertanian Olah Tanah Konservasi, Gorontalo, 6–7 October 2004, pp 18–35 (in Indonesian)

Utomo M, Suprapto H, Sunyoto (1989) Influence of tillage and nitrogen fertilization on soil nitrogen, decomposition of alang-alang (*Imperata cylindrica*) and corn production of alang- alang land. In: Heide J van der (ed) Nutrient management for food crop production in tropical farming systems. Institute for Soil Fertility (IB), Haren (Gr.), the Netherlands, pp 367–373

Utomo M, Niswati A, Dermiyati WMR, Raguan AF, Syarif S (2010a) Earthworm and soil carbon sequestration after twenty one years of continuous no-tillage corn-legume rotation in Indonesia. J Int Fuzzy Syst 7:51–58

Utomo M, Buchari H, Banuwa IS (2010b) Peran olah tanah konservasi jangka panjang dalam mitigasi pemansan global: penyerapan karbon, pengurangan gas rumah kaca dan peningkatan produktivitas lahan. Laporan Akhir Hibah Kompetitif penelitian sesuai prioritas nasional. Tahun Kedua. DP2M (in Indonesian, unpublished)

Utomo M, Buchari H, Banuwa IS (2011) Peran olah tanah konservasi jangka panjang dalam mitigasi pemansan global: penyerapan karbon, pengurangan gas rumah kaca dan peningkatan produktivitas lahan. Laporan Akhir Hibah Kompetitif penelitian sesuai prioritas nasional. Tahun Kedua. DP2M (in Indonesian, unpublished)

Utomo M, Buchari H, Banuwa IS, Fernando LK (2012) Carbon storage and carbon dioxide emission as influenced by long-term conservation tillage and nitrogen fertilization in corn–soybean rotation. J Trop Soils 17(1):75–84

Utomo M, Banuwa IS, Buchari H, Anggraini Y, Berthiria (2013) Long-term tillage and nitrogen fertilization effects on soil properties and crop yields. J Trop Soils 18(20):21–30

Wright AL, Hons FM (2004) Soil aggregation and carbon and nitrogen storage under soybean cropping sequences. Soil Sci Soc Am J 68:507–513

# Chapter 5
# Improving Biodiversity in Rice Paddy Fields to Promote Land Sustainability

**Dermiyati and Ainin Niswati**

**Abstract** Rice is a staple food for many people in the world, especially in Asian countries, and rice consumption increases every year. Efforts have been made to increase rice production, leading to social, economic, and environmental impacts. Rice in paddy fields is mostly grown using conventional farming systems with high inputs of agrochemicals (inorganic fertilizers and chemical pesticides). Continuous application of agrochemicals may damage the soil and cause decreased soil productivity and biodiversity, as well as increased pest attacks and methane emissions. Therefore, organic farming systems are likely to be the best practices for promoting land sustainability. In fact, farmers in many countries have shifted their rice production management from conventional to organic. However, it is argued that, in general, organic systems are related to lower yields and lower environmental impacts while conventional systems are related to higher yields and higher environmental impacts. Although the movement from conventional to organic farming systems is believed to have positive short-term impacts by improving soil biodiversity, therefore, it will also have an impact in terms of lowering rice production volumes. The achievement of food security and food availability requires government policies to promote the use of organic fertilizers and subsidize their prices, as well as regulation to support high prices for organic products. Application of organic fertilizers and biofertilizers, and the use of crop rotation, are likely to improve soil fertility, which is related to increased biodiversity, and eventually this will contribute to higher rice production volumes in the long term.

**Keywords** Biodiversity in rice paddy fields • Conventional and organic farming • Soil biodiversity

Dermiyati (✉) • A. Niswati
Soil Science Department, Faculty of Agriculture, University of Lampung,
Jl. Sumantri Brojonegoro No.1, Bandar Lampung 35145, Indonesia
e-mail: dermiyati@yahoo.co.id; dermiyati.1963@fp.unila.ac.id;
ainin.niswati@fp.unila.ac.id; niswati@yahoo.com

N. Kaneko et al. (eds.), *Sustainable Living with Environmental Risks*,
DOI 10.1007/978-4-431-54804-1_5, 

## 5.1 Introduction

Various types of organisms live in wetland rice and each of them interacts with others to form a specific food chain (Ali 1990; Roger et al. 1993). The direct or indirect effect of one organism on another will affect the community structure of the wetland rice organisms, which consist of nematodes, microcrustacea, protozoa, insect larvae, algae, mollusca, and oligochaetae (Mogi 1993). As the organisms interact with each other, their populations are also affected by environmental factors, including fertilizer and pesticide application, water management, and crop variety (Simpson et al. 1994).

The role and potential of microorganisms and invertebrates in biodiversity and sustainability of wetland rice production has been reviewed by Roger et al. (1991). Suitable rice-producing environments are essential for wetland rice production. It is therefore important to consider microorganisms and invertebrates as well as their biodiversity in wetland rice. Sustainable rice-producing environments depend on microbial and invertebrate populations, agricultural practices, the status of germplasm collection, and developments in biotechnology.

Moreover, whilst crop intensification using agrochemicals does increase yields, on the other hand it also reduces the number of edible species traditionally harvested from ricefields, such as snails, prawns, crabs, large water bugs, fish, and frogs (Heckman 1979). Agrochemicals also cause uncontrolled growth of single species that might, directly or indirectly, have detrimental effects, such as the outbreak of pests (Heinrichs 1988). In addition, fertility- or health-related aspects of ecosystems may be affected by other organisms, such as: (1) blooms of unicellular algae (observed after fertilizer application, which causes nitrogen losses by volatilization); (2) proliferation of ostracods and chironomid larvae (observed after insecticide application, which inhibits the development of efficient nitrogen-fixing blue-green cyanobacterial blooms); and (3) proliferation of snails or mosquito larvae (observed after insecticide application, which causes vector-borne diseases) (Roger and Kurihara 1988).

Nowadays, rice is generally cultivated in two forms: conventional and organic. In Indonesia, however, some farmers choose to follow a middle course, known as "semi-organic." Many reports suggest that organic agriculture is more efficient, and also effective in reducing water and soil pollution (Erhart and Hartl 2009), greenhouse gas emissions (Lumbaraja et al. 1998), and risks to human health (Mader et al. 2002), as well as increasing energy efficiency (Mansoori et al. 2012). Research was conducted in Iran on the relative energy efficiency and economic benefits of organic versus conventional rice production. The organic farms performed better than conventional rice production systems in terms of all energy efficiency indexes, as well as cost-to-benefit ratios and gross and net returns, while total costs of production were also lower (Mansoori et al. 2012). Previously, Lumbaraja et al. (1998) had studied methane emissions from Indonesian rice fields in several locations in Sumatra and Bali. They found that methane emissions in rain-fed conditions were 27–37 % lower than in continuously flooded conditions. In separate research, it was noted that paddy fields contribute about 10 % of all global methane emissions (Oyewole 2012).

This chapter considers the sustainability of rice production when shifting from conventional to organic farming systems according to Indonesian experiences. It focuses especially on paddy fields containing microbial and invertebrate populations.

## 5.2 Indonesian Experiences: Effects on Soil and Water Biodiversity of Shifting from Conventional to Organic Farming in Paddy Fields

Most rice field ecosystems in Indonesia are managed using either conventional or organic methods, or a combination of the two. Conventional management involves applying agrochemicals such as inorganic fertilizers and chemical pesticides. Recently, however, the government is socializing organic or semi-organic management methods among farmers' groups. Any resulting change from conventional to organic farming systems is likely to affect biotic environments in rice paddy fields. Although changes in biodiversity are due to the different inputs for conventional farming as opposed to organic, the extent to which inputs affect biodiversity needs to be clarified.

We conducted research during 2006–2007 in the Pagelaran subdistrict of the Tanggamus district in Lampung Province. The research studied how the physical, chemical, and biological properties of soil altered as a result of the farmers changing their rice paddy fields from conventional to organic cultivation. In the following description, however, we will focus only on changes affecting soil and water biodiversity in the paddy fields.

The farmers had cultivated rice in the paddy fields since the 1970s and had used high inputs of agrochemicals, high quality seeds, etc., to generate higher rice production volumes. From the year 2000, however, some farmers who were members of *Ikatan Petani Pengendali Hama Terpadu* (IPPHT, the Farmers' Association for Integrated Pest Management) realized that although they were using high inputs the rice yields were not increasing anymore and the soil fertility had deteriorated. The soil was hard, and cracking, and lacking in fertility due to reduced levels of organic matter. From then on, therefore, certain farmers pioneered the use of an organic fertilizer called "*bokashi*." *Bokashi* was introduced by Japanese farmers; it is a compost made from paddy husk, cow manure, and microorganisms that act as decomposers. Recently it has become very popular in Indonesia too. The farmers make their own *bokashi* using microorganisms produced using local materials (i.e., papaya fruits are mixed with coconut water and palm sugar, then fermented for 2 weeks, and the supernatant which contains microbes is used as decomposer).

The Pagelaran farmers refer to themselves as organic paddy farmers and they cultivate rice in paddy fields twice a year (there are two seasonal planting periods, in the wet and dry seasons). During each planting period they apply about 4 t *bokashi* ha$^{-1}$ and they do not use chemical fertilizers anymore. They also no longer use chemical pesticides, which they replaced with organic pesticides made from local plants such as tobacco and ginger.

Our study commenced in 2006, and we surveyed farmers who converted to organic methods during the period from 2000 to 2005, since each farmer converted at a different time. Within the sample, therefore, farmers who started organic farming in 2005 had one year's application of *bokashi*, those who started in 2004 had 2 years of application, and so on. Farmers who had not applied *bokashi* by 2006 (non-organic farmers) were used as a control (0 years' application). Some farmers did not want to be organic farmers because they were afraid their rice production would be low if they did not use agrochemicals as fertilizers and pesticides. The results of the study are reviewed below.

### 5.2.1 Effects on Water Organisms in Paddy Fields

Aquatic biodiversity in rice fields is endangered as a result of the expansion of human populations. It has led to increasing pressure on living aquatic resources in rice fields due to agrochemical use and runoff, sedimentation, habitat loss, destruction of fish breeding grounds, and unsustainable fishing methods (Halwart 2004).

Information about the types of organism living in paddy fields is necessary because they can serve as indicators for environmental risks. Niswati and Purnomo (2007) studied how the community structure, diversity, and population density of aquatic organisms in Lampung Province was affected by whether paddy fields were conventional or organic (Table 5.1). Their study found 22 types of water organism with a size of 50 μm to 1 cm in the four types of study site. The study sites were categorized according to farming system and location, as follows: (1) Conventional, Taman Bogo; (2) Organic, Pagelaran; (3) Conventional, Pagelaran; and (4) Organic, Greenhouse. Organisms from the groups of Cladocera, Cyclopoida, Ploimida, Zygnemetales, Nematoda, Diptera, Podocopida, Volvocida, and Archipora were found in all locations. However, Anostraca, Ephemeraptera, Closterium, Bdelloida, and Haplatoxida were found only in the conventional paddy fields in Taman Bogo, while a greater total number of organisms was found in Taman Bogo's flooded paddy fields than in Pagelaran. In Pagelaran there were a greater variety of organism types in the organic paddy fields than in the conventional fields. Meanwhile, organism populations in the organic paddy field inside the greenhouse were higher than in the external paddy fields but the organisms were more homogenous.

Organism population changes in the conventional paddy fields in Taman Bogo and Pagelaran, as well as the organic paddy fields in Pagelaran showed that Cladocera, Cyclopoida, Ploimida, and Volvocida are found throughout the duration of plant growth, while other organisms are not. Some organisms, namely Nematodes, Ephemeraptera, and Chlorococcum, are found only at the beginning of plant growth; on the other hand, Bdelloida, Turbellaria, and Archipora are found only at the end

**Table 5.1** Types of water organisms in Lampung Province and their abundances (individual $m^{-2}$) (Niswati and Purnomo 2007)

| | Farming system types/locations | | | |
|---|---|---|---|---|
| Taxonomic groups | Conventional, Taman Bogo[a] | Organic, Pagelaran[b] | Conventional, Pagelaran[b] | Organic, Greenhouse[c] |
| Cladocera | 56,608 | 27,520 | 1,152 | 142,500 |
| Cyclopoida | 127,384 | 93,824 | 33,856 | 3,500 |
| Ploimida | 24,192 | 33,600 | 21,88 | 25,250 |
| Zygnemetales | 12,416 | 6,400 | 3,392 | 14,750 |
| Anostraca | 824 | 0 | 0 | 0 |
| Nematoda | 752 | 6,080 | 13,632 | 2,250 |
| Algae spyrogira | 1,288 | 768 | 448 | 0 |
| Diptera | 360 | 1,600 | 1,472 | 4,000 |
| Podocopida | 2,496 | 2,880 | 2,688 | 6,000 |
| Ephemeraptera | 280 | 0 | 0 | 0 |
| Closterium | 688 | 0 | 0 | 0 |
| Bugs | 1,064 | 64 | 0 | 0 |
| Isopoda | 584 | 0 | 192 | 0 |
| Chlorococcum | 1,240 | 0 | 0 | 11,500 |
| Volvocida | 5,280 | 640 | 1,344 | 19,500 |
| Bdelloida | 200 | 0 | 0 | 0 |
| Haplatoxida | 216 | 0 | 0 | 0 |
| Turbellaria | 480 | 3,328 | 1,088 | 10,750 |
| Paramecium | 0 | 128 | 576 | 40,750 |
| Euglenida | 0 | 256 | 1,408 | 82,500 |
| Archipora | 1,158 | 1,280 | 0 | 0 |
| Sessilida | 0 | 192 | 128 | 0 |

[a] Seven observations
[b] Five observations
[c] Ten observations

of plant growth. Meanwhile, Podocopida, Algae Spyrogira, Zygnemetales, Diptera, Closterium, Bugs, and Haplatoxida were found inconsistently. This type of succession was similar to that reported by Yamazaki et al. (2004).

In comparison, surveys conducted in Sri Lanka on biodiversity in wetland rice field ecosystems documented 494 species of invertebrates belonging to 10 phyla, and 103 species of vertebrates, while the flora included 89 species of macrophytes, 39 genera of microphytes, and 3 species of macrofungi (Bambaradeniya et al. 2004).

Moreover, changes affecting dominant protozoa and algae populations in conventional and organic paddy field flooding water in Lampung Province were also studied (Niswati et al. 2008). There were two genera of protozoa (*Euglena* sp. and *Pleodorina* sp.) and two genera of algae (*Volvox* sp. and *Diatom*) that were dominant in the paddy fields where *bokashi* was continuously applied (Fig. 5.1). Among dominant protozoa, the population of *Volvox* sp. was significantly influenced by the continuous application of *bokashi*. The populations of protozoa and algae were higher under continuous *bokashi* application (for 2–4 years) than under the control.

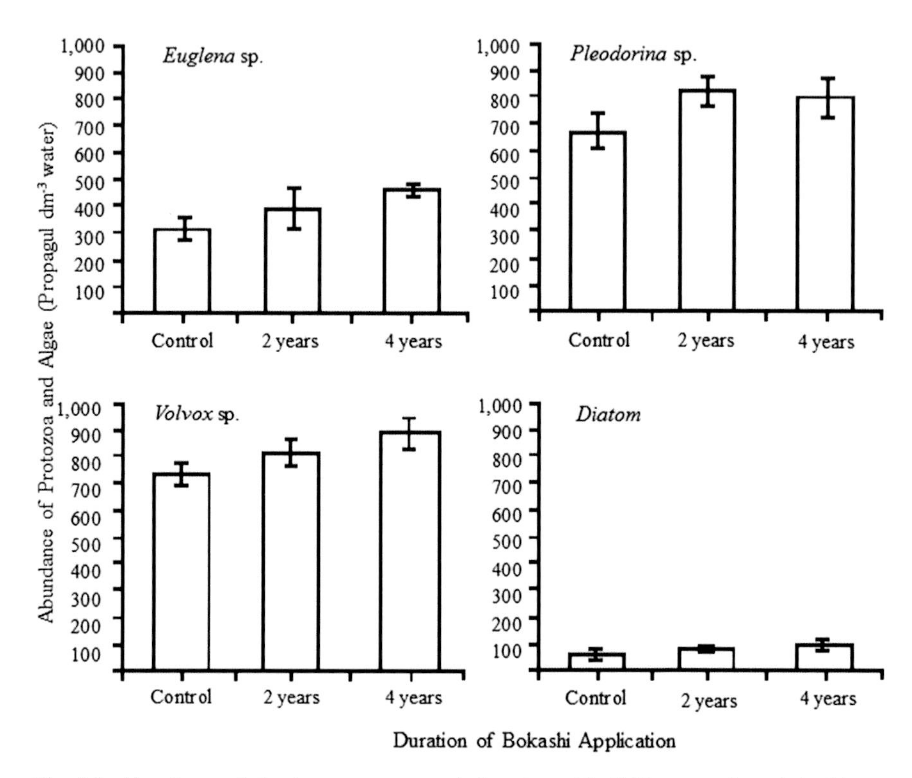


**Fig. 5.1** Abundance of dominant protozoa and algae in paddy field water where *bokashi* was applied continuously. *Bars* indicate standard error (P = 0.95) (Niswati et al. 2008)

Fluctuation in numbers of protozoa and total algae in organic paddy fields was higher than under conventional cultivation systems, the highest being after 4 years of *bokashi* application. Meanwhile, other protozoa and algae were also found, namely Chlorococcum, Archipora, Bdelloida, Algae Spyrogira, and Ploimida, but they were not dominant.

Populations of dominant protozoa and algae were likely to increase, starting 30 days after planting and continuing steadily until harvesting time (data not shown). This is because continuous application of *bokashi* compost may increase the populations of bacteria and fungi in the soil (Labidi et al. 2007) as well as increasing carbon biomass, nitrogen, phosphorus, and sulfur. Bacteria and fungi can act as food sources for protozoa, so continuous application of *bokashi* may increase microbial activity in the soil, which may increase the protozoa population. The decreases in protozoa and algae populations that occurred at 30 and 60 days after planting time were due to organic paddy field management involving weed cutting, wetting by irrigation, and drying out of the water in the paddy fields.

The biodiversity index for protozoa and algae was the same for both conventional and organic paddy fields. It is likely that the application of *bokashi* compost affected all organisms in the flooding water of the paddy fields so that the food chain was not yet affected.

In separate research, Roger et al. (1991) summarized some studies from multiple countries about species abundance in traditional rice fields. They reported that based on a 1975 study in Thailand the species abundance in one traditional rice field in 1 year was 590 species (excluding fungi) (Heckman 1979). Moreover, about 39 taxa of aquatic invertebrates were reported following a 2-year study of pesticide application on Malaysian rice fields (Lim 1980). Across 18 sites in the Philippines and India the highest number of aquatic invertebrate taxa reported was 26 and the lowest was 2 by single sampling at individual sites (Roger et al. 1987). Based on these data recorded from 1975 to the present, they stated that crop intensification had a tendency to decrease the values for total number of species; however, it cannot be accepted as a general concept that crop intensification decreases biodiversity in rice fields (Simpson et al. 1994).

### 5.2.2 *Effects on Soil Microorganisms in Paddy Fields*

The soil microbial community is involved in numerous ecosystem functions, such as nutrient cycling and organic matter decomposition, and plays a crucial role in the terrestrial carbon cycle (Schimel 1995). Changes in populations of phosphate solubilizing microorganisms in conventional and organic paddy fields were also studied (Dermiyati et al. 2009). Although the populations of phosphate solubilizing microorganisms were not affected by continuous application of *bokashi* and the contribution to soil P from *bokashi* was relatively low, the microorganisms did play a role in the availability of soil available-P from residual P fertilizers that were applied intensively for long periods. Figure 5.2 shows the changes in phosphate solubilizing microorganism populations as a result of converting from conventional to organic paddy fields by applying *bokashi* continuously.

Roger et al. (1991) also reviewed effects of crop intensification on soil and water microbial populations. The impacts of crop intensification on the rice field microflora due to pesticide use are: (1) alteration of activities related to soil fertility, and (2) reduction of pesticide efficiency because of shifts in microbial populations toward organisms more efficient in their degradation.

## 5.3 Strategies to Improve Biodiversity in Rice Paddy Fields

There are measures that can be taken to improve biodiversity in rice paddy fields, such as continuous application of organic fertilizers. Alternatively, biofertilizers can be applied to introduce microorganisms that benefit nutrient cycles and energy supply. Farmers can also undertake crop rotation to cut the life cycles of pests.

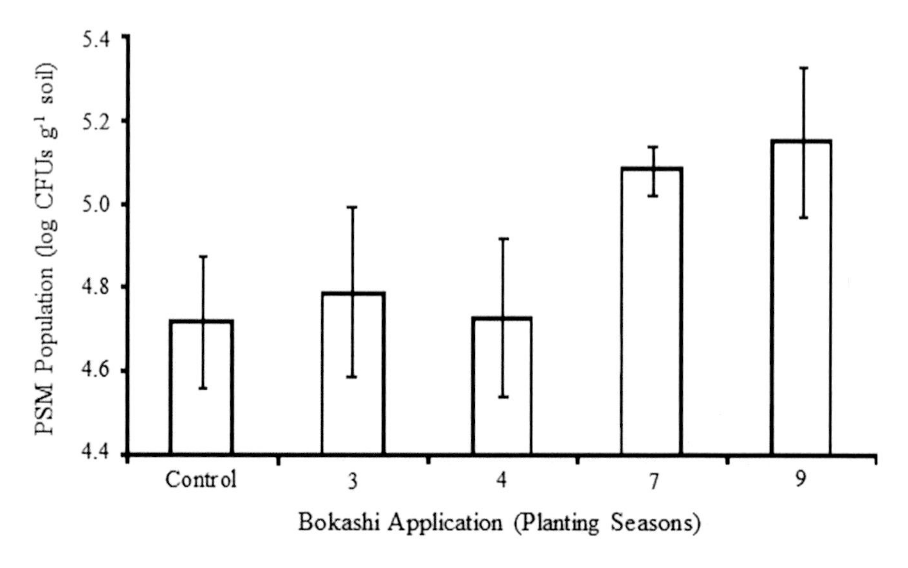


**Fig. 5.2** Populations of phosphate solubilizing microorganisms (PSM) as a result of conversion from conventional to organic paddy fields by applying *bokashi* continuously in the Pagelaran subdistrict. *Bars* indicate standard error (P = 0.95) (Dermiyati et al. 2008)

### *5.3.1 Application of Organic Matter and Biofertilizers*

In theory, addition of organic matter promotes biological and microbial activities that accelerate the breakdown of organic substances in the organic matter. The use of organic materials such as animal manure, crop residues, green manure, and *bokashi* compost as alternative sources is promising.

Soil microbial biomass comprises living plant roots and organisms, and the living portion of soil organic matter. It acts as the agent of biochemical changes in soil and as a repository of plant nutrients such as nitrogen (N) and P in agricultural ecosystems (Jenkinson and Ladd 1981). Lower microbial biomass in soils from conventional agro-ecosystems is often caused by reduced organic carbon content in the soil (Fliebach and Mader 2000). The quantity and quality of organic inputs are the most important factors affecting microbial biomass and community structure (Peacock et al. 2001). Continuous cultivation with frequent tillage results in rapid loss of organic matter through increased microbial activity (Shepherd et al. 2001).

Nakhro and Dkhar (2010) observed that the application of organic fertilizers increased fungal and bacterial populations as well as microbial biomass carbon compared to the application of inorganic fertilizers or a control. The increases were found both at surface soil depth (0–15 cm) and at sub-surface soil depth (15–30 cm); however, the increases were greater at sub-surface soil depth. This could be due to the addition of organic amendments that might have a large impact on soil microbial activity (Elliot and Lynch 1994), microbial diversity (Girvan et al. 2004), and bacteria density (Bruggen-van and Semenov 2000).

### 5.3.2 Crop Rotation

The main components of organic farming systems are soil protection, bio-control, nutrient cycles, and biodiversity. So far, farmers' efforts have been limited to replacing the use of agrochemicals (fertilizers and pesticides) with organic fertilizers (such as compost) and pesticides. However, this approach primarily addresses soil protection, and only partly influences bio-control and nutrient cycles, while no attention has yet been focused on biodiversity. Although monoculture is part of the "Green Revolution," or conventional, farming systems, it still prevails in paddy fields that have been declared organic. In order to improve biodiversity, therefore, crop rotation is necessary in these paddy fields.

Crop rotation affects microbial populations. Research in the Pagelaran district showed that there were differences in the populations of phosphate solubilizing microbes in paddy fields where *bokashi* compost was continuously applied. Crop rotation from paddy to legumes caused a decrease in the population of phosphate solubilizing microbes (Dermiyati et al. 2009). It was likely that the cultivation of legumes increased the nitrogen-fixing bacteria rhizobium. These bacteria have an ability to fix nitrogen from the air and make it available for the plant by cooperation with legume roots. Moreover, it was observed that the activity of phosphate solubilizing bacteria in the rhizosphere was affected by the presence of nitrogen- fixing bacteria like rhizobium (Widawati 1999).

## 5.4 Conclusions

As mentioned previously, improvements are necessary in rice paddy fields where heavy chemicals have been applied so that biodiversity can be maintained and sustainability in wetland rice growing is achieved. Shifting from conventional to organic farming needs a strong policy from the government to make organic fertilizers available at prices farmers can afford, or to subsidize their prices for the farmers. Government policy has to encourage and incentivize farmers who conduct organic farming and support high prices for organic products. Meanwhile, the farmers themselves must change their habits and culture with regard to paddy field cultivation to increase yields for food security. In relation to food security, production of organic fertilizers that contain high levels of nutrients is recommended. In addition, agricultural extension must teach the farmers simple technologies such as how to make *bokashi* or compost using local materials available in the villages.

To establish organic agriculture as an important tool in sustainable food production, assessments of the many social, environmental, and economic benefits of organic farming systems need to be complemented by a fuller understanding of the factors limiting organic yields. Agro-ecologists and conservation biologists should work together to formulate strategies based on biodiversity as an organizing principle in the sustainable management of rice field agro-ecosystems.

## References

Ali AB (1990) Seasonal dynamics of microcrustacean and rotifer communities in Malaysian rice fields used for rice-fish farming. Hydrobiologia 206:139–148

Bambaradeniya CNB, Edirisinghe JP, De Silva DN, Gunatilleke CVS, Ranawana KB, Wijekoon S (2004) Biodiversity associated with an irrigated rice agro-ecosystem in Sri Lanka. Biodivers Conserv 13(9):1715–1753

Bruggen-van AHC, Semenov AM (2000) In search of biological indicators for soil health and disease suppression. Appl Soil Ecol 15:13–24

Dermiyati AJ, Yusnaini S, Nugroho SG (2009) Change of phosphate solubilizing bacteria population on paddy field with intensive farming became sustainable organic farming system. J Trop Soils 14(2):143–148 (in Indonesian)

Elliot LF, Lynch JM (1994) Biodiversity and soil resilience. In: Greenland DJ, Szabolcs I (eds) Soil resilience and sustainable land use. CAB International, Wallingford, pp 353–364

Erhart E, Hartl W (2009) Soil protection through organic farming: a review. In: Lichtfouse E (ed) Organic farming, pest control and remediation of soil pollutants. Sustain Agric Rev 1: 203–226

Fliebach A, Mader P (2000) Microbial biomass and size-density fractions differ between soils of organic and conventional agricultural systems. Soil Biol Biochem 32:757–768

Girvan MS, Bullimore J, Ball AS, Pretty JN, Osborn AM (2004) Response of active bacterial and fungal communities in soil under winter wheat to different fertilizer and pesticide regimens. Appl Environ Microbiol 70:2692–2710

Halwart M (2004) Aquatic biodiversity in ricefields. International year of rice: rice is Life. www.rice2004.org

Heckman CW (1979) Ricefield Ecology in Northeastern Thailand. (Monographiae Biologicae). Dr. W. Junk bv Publishers, Hague, p 228

Heinrichs EA (1988) Role of insect-resistant varieties in rice IPM systems. In: Teny PS, Heong KL (eds) Pesticide management and integrated pest management in Southeast Asia. International Crop Protection, Bettsville

Jenkinson DS, Ladd JN (1981) Microbial biomass in soil: measurement and turnover. In: Paul EA, Ladd JN (eds) Soil biochemistry. Marcel Dekker, New York, pp 415–471

Labidi S, Nasr H, Zouaghi M, Wallander H (2007) Effects of compost addition on extra-radical growth of arbuscular mycorrhizal fungi in *Acacia tortilis* sp. raddiana savanna in a pre-Saharan area. Appl Soil Ecol 35:184–192

Lim RP (1980) Population changes of some aquatic invertebrates in ricefields. In: Tropical ecology and development. Proceedings of the 5th international symposium of tropical ecology. International Society of Tropical Ecology, Kuala Lumpur, pp 971–980

Lumbaraja J, Nugroho SG, Niswati A, Ardjasa WS, Subadiyasa N, Arya N, Haraguchi H, Kimura M (1998) Methane emission from Indonesian ricefields with special references to the effect of yearly and seasonal variations, rice variety, soil type and water management. Hydrolog Process 12:2057–2072

Mader O, Flie-Bach A, Gunst L, Fried P, Niggli U (2002) Soil fertility and biodiversity in organic farming. Science 296:1694–1697

Mansoori H, Moghaddam PR, Mohadi R (2012) Energy budget and economic analysis in conventional and organic rice production systems and organic scenarios in the transition period in Iran. Front Energ 6(4):341–350. doi:10.1007/s11708-012-0206-x

Mogi M (1993) Effect of intermittent irrigation on mosquitoes (Diptera: Culicidae) and larvivorous predators in rice fields. J Med Entomol 30:309–319
Nakhro N, Dkhar MS (2010) Impact of organic and inorganic fertilizers on microbial populations and biomass carbon in paddy field soil. J Agron 9(3):102–110
Niswati A, Purnomo (2007) The changes of communities and diversity of aquatic organisms on the floodwater of paddy fields of Pagelaran and Taman Bogo, Lampung Province. J Akta Agrosia (Special Edition) 2:213–219 (in Indonesian)
Niswati A, Dermiyati, Arif MAS (2008) Changes in the dominant population of protozoa and algae inhabiting the floodwater of paddy fields subjected to continued *bokashi* application. J Trop Soils 13(3):225–231 (in Indonesian)
Oyewole OA (2012) Microbial communities and their activities in paddy fields: a review. J Vet Adv 2(2):74–80
Peacock AD, Mullen MD, Ringelberg DB, Tyler DD, Hedrick DB, Gale PM, White DC (2001) Soil microbial community responses to dairy manure or ammonium nitrate applications. Soil Biol Biochem 33:1011–1019
Roger PA, Kurihara Y (1988) Floodwater biology of tropical wetland rice fields. In: Proceedings of the first international symposium on paddy soil fertility, University of Chiang Mai, Thailand, 6–13 December 1988, pp 275–300
Roger PA, Santiago-Ardales S, Watanabe I (1987) The abundance of heterocystous blue-green algae in rice soils and inocula used for application in rice fields. Biol Fertil Soils 5:96–105
Roger PA, Heong KL, Teng PS (1991) Biodiversity and sustainability of wetland rice production: role and potential of microorganisms and invertebrates. In: Hawksworth DL (ed) The biodiversity of microorganisms and invertebrates: its role in sustainable agriculture, CAB International, Wallingford, England, pp 117–136
Roger PA, Zimmerman WJ, Lumpkin TA (1993) Microbial management of wetland rice fields. In: Metting FB Jr (ed) Soil microbial ecology: applications in agricultural and environmental management. Marcel Dekker, Inc., New York, pp 417–455
Schimel DS (1995) Terrestrial ecosystem and carbon cycle. Global Change Biol 1:77–91
Shepherd TG, Sagar S, Newman RH, Ross CW, Dando JL (2001) Tillage-induced changes in soil structure and soil organic matter fractions. Aust J Soil Res 39:465–489
Simpson IC, Roger PA, Oficial R, Grant IF (1994) Effects of nitrogen fertilizer and pesticide management of floodwater ecology in a wetland rice field II. Dynamics of microcrustaceans and dipteran larvae. Biol Fertil Soils 17:138–146
Widawati S (1999) Effect of soil microbe application on the growth and yields of soybean (*Glycine max* L.) in an acid soil. J Microbiol Trop II(2):61–67 (in Indonesian)
Yamazaki M, Hamada Y, Kamimoto M, Momii T, Kimura M (2004) Composition and structure of aquatic organism communities in various water conditions of a paddy field. Ecol Res 19:645–653

# Chapter 6
# Agroforestry Models for Promoting Effective Risk Management and Building Sustainable Communities

**Damasa B. Magcale-Macandog**

**Abstract** Soil erosion and environmental degradation due to the cultivation of marginal upland areas are now considered major environmental risks in the Philippines. Agroforestry may help address the situation. In agroforestry systems, the positive interactions of tree-crop combinations not only improve biophysical conditions in farms, but also enhance food security in farming households.

A combination of Participatory Rural Appraisal (PRA), a household survey, focus group discussions, field experiments, and simulation modeling was undertaken in Claveria, Misamis Oriental, Philippines. The agroforestry system adopted depended on the farmers' motivations. The adoption of agroforestry significantly increased the households' level of income by around 42–137 %, compared with that from continuous annual mono-cropping. Another beneficial feature of an agroforestry system was the enhanced nutrient inflow to the system through leaf litterfall, stemflow, and throughfall. A modeling study using the WaNuLCAS model showed that the *Eucalyptus*-maize hedgerow system provided significant improvements to a range of biophysical and economic measures of productivity and sustainability.

It is recommended that both national and local government units mainstream their policies and efforts toward promoting agroforestry adoption in the Philippine uplands.

**Keywords** Agroforestry • Food security • Land degradation • Leaf litterfall • Nutrient inflow • Soil erosion

D.B. Magcale-Macandog (✉)
Institute of Biological Sciences, University of the Philippines Los Baños, College Laguna 4031, Philippines
e-mail: demi_macandog@yahoo.com

N. Kaneko et al. (eds.), *Sustainable Living with Environmental Risks*,
DOI 10.1007/978-4-431-54804-1_6, © The Author(s) 2014

## 6.1 Introduction

The uplands in the Philippines are of great importance and interest because they comprise about 59 % of the country's total land area. They are dynamic and highly interactive landscape components of the rural system, and also serve as the life support for the lowlands and coastal areas. In addition, they are home to the increasing population of the "poorest of the poor," and are expected to absorb more of the expanding population (Sajise and Ganapin 1991).

The Philippine uplands are a very heterogeneous and fragile resource base (Sajise and Ganapin 1991). Most of these areas are either open grassland, degraded, or occupied by settlers (Villancio et al. 2003). More than 20 million people are estimated to have settled in the uplands, and the number is increasing at a rate of about 2.8 % annually, which is above the national average of 2.32 %.

Its geographical location has made the Philippines highly vulnerable to natural hazards, the most common of which is the occurrence of turbulent typhoons. During the decade from 2001 to 2012, the country was hit by a total of 184 typhoons, or an average of 18 typhoons per year (Israel and Briones 2012). Climate change is perceived to have increased the frequency and intensity of heavy rainfall associated with typhoons and other weather systems, resulting in flooding. From 2000 to 2010, damage to agricultural crops caused by typhoons, floods, and droughts amounted to nearly PHP 106.88 million. Rice, corn, and other high value cash crops sustained the most damage. Typhoons Ondoy (Ketsana) and Pepeng (Parma) in September and October 2009 wrought havoc in both urban and rural areas in the country, with total damage reaching PHP 36.2 billion.

In the uplands, a major problem is food insecurity, which is mainly a consequence of land degradation. There is general recognition of the serious implications of deforestation, soil erosion, declining agricultural productivity, loss of biodiversity, off-site impacts, increasing poverty, and the social costs associated with the biophysical and ecological instability in the uplands. While 53 % of the Philippines' total land area is classified as forestlands, only 17 % is adequately covered with forest vegetation. In fact, the total forest cover in the country declined by as much as 3.54 % for the period 1990–1995, the fourth highest loss rate in the world. This rapid decline in forest areas can be attributed to the large and rapid conversion of the Philippine uplands into permanent annual cropping areas to meet the food requirements of an increasingly expanding population (Domingo and Buenaseda 2000). However, the productivity of sloping lands has been diminishing at an alarming rate due to soil degradation or erosion brought about by the activities of this population as it grows. According to Escaño and Tababa (1998), the rates of soil erosion in sloping areas range from 23 to 218 ton/ha/year for bare plots on gradients of 27–29 % to 36–200 ton/ha/year on plots cultivated up and down the hill. These rates are higher than the acceptable soil loss level of 3–10 ton/ha/year (Paningbatan 1989), and the situation poses a grave threat to the productivity and sustainability of farming in the upland areas.

In summary, the uplands can be characterized as degraded and ecologically marginal for agricultural purposes with landscapes that are highly sensitive and of low resilience. The biophysical limitations of these lands affect production, income, and household food security. Diminished food access due to the degraded natural resources, higher food prices, limited income opportunities, and the impact of natural elements leave the upland population a legacy of poverty and food insecurity.

Agroforestry is a dynamic, ecologically-based, natural resource management system that, through the integration of trees into farms, diversifies and sustains smallholder production for increased social, economic, and environmental benefits (Leaky 1996). Introducing trees within the cropping system can help prevent land degradation, increase biodiversity, and at the same time allow the continued use of the land for agricultural crop production (Wise and Cacho 2002).

Mature trees in agroforestry systems can yield numerous positive effects on cropped fields (Garcia-Barrios and Ong 2004). Among these are improved soil fertility and physical properties via organic matter addition from litter; reduced soil erosion through stabilization of loose soil surface by tree roots; recovery of leached nutrients from deep soil layers inaccessible to crops; reduced soil evaporation, leaf temperature, and evaporative demand by crops via tree shade; increased soil infiltration rate; protection against wind and runoff; reduced weed population; and reduction and potential slowdown of windborne pests and diseases. In a system where nitrogen (N)-fixing trees are used as hedgerows, alternative sources of N for trees can significantly reduce competition with crops.

## 6.2 Agroforestry Adoption, Innovations, and Smallholder Farmers' Motivations in Claveria, Misamis Oriental

### *6.2.1 Description of the Study Site*

Claveria is a land-locked agricultural municipality in the province of Misamis Oriental in northern Mindanao (Fig. 6.1). It is a volcanic plateau ascending abruptly from about 350 masl (meters above sea-level) in the west to about 1,200 masl in the east. Its topography is generally rugged, characterized by gently rolling hills and mountains with cliffs and escarpments. More than 68 % of Claveria's total land area has slopes greater than 18 %.

The soil in Claveria is classified as Jasaan Clay, with a deep soil profile (>1 m) and rapid drainage (Bureau of Soils 1985). The soil is generally acidic (pH 3.9–5.2), with low cation exchange capacity (CEC), and a low to moderate organic matter content (1.8 %). It also has low levels of available phosphorus and exchangeable pottasium (Magbanua and Garrity 1990).

The climate of Claveria is classified as having a rainfall distribution of 5 or 6 wet months (>200 mm/month) and 2 or 3 dry months (<100 mm/month). However, rainfall patterns throughout the municipality vary with elevation, with the upper areas having a relatively greater amount of rainfall than the lower areas.

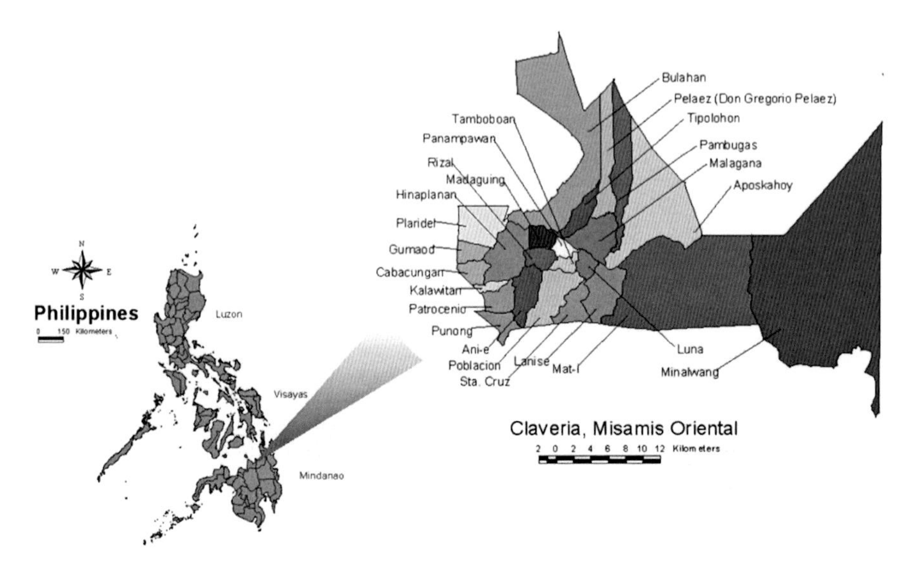


**Fig. 6.1** Claveria, Misamis Oriental, Philippines, including village administrative boundaries

### 6.2.2 *Participatory Rural Appraisal (PRA) and Household Survey*

A 4-day PRA with key informants from 17 *barangays* (villages) in Claveria was conducted to obtain a preliminary assessment of the basic and universal information held by the community on the issues to be addressed (Magcale-Macandog et al. 2006). Three hundred farmer-respondents were selected for the household survey using a stratified sampling technique whereby the total population of farmers in the study area was divided into subpopulations (strata) based on elevation (Upper, Middle, and Lower Claveria) and the type of agroforestry or non-agroforestry system practiced (Magcale-Macandog et al. 2010). The number of random samples representing each stratum was proportional to the size of the subpopulation of farmers practicing the type of agroforestry system at each elevation class. Farmers practicing agroforestry systems were classified based on the spatial arrangements of trees in their systems (Fig. 6.2) as follows: (1) parkland planting (scattered trees), (2) hedgerow system (planting of trees along farm contours), (3) block planting or *taungya* system, and (4) multi-story system.

### 6.2.3 *Drivers of Land Degradation in Claveria*

Farmers identified the poor growth of plants or grasses and low crop yield as the main indicators of soil degradation. Degraded soils were also characterized by acidic or red soil, absence of earthworms, faster drying up of soil, and absence of trees.

**Fig. 6.2** Common farming systems adopted in Claveria, Misamis Oriental, Philippines: (**a**) parkland system; (**b**) block planting with timber trees; (**c**) block planting with fruit trees; (**d**) natural vegetative strips (NVS); (**e**) coffee-based hedgerow intercropping; (**f**) timber-based hedgerow intercropping; (**g**) border planting with fruit trees, (**h**) border planting with *Gliricidia sepium*; (**i**) home garden

In Lower Claveria, continuous land cultivation (25 %) and fertilizer application (25 %) were perceived as the main factors that led to soil degradation. In Upper Claveria, however, 67 % of the farmers said that soil erosion was the main cause of soil degradation. In both areas, farmers associated soil degradation with plowing along the contour, not observing proper soil conservation, burning farm waste, and pesticide application.

Farmers thought that contouring was the primary strategy to address soil degradation. Planting trees or agroforestry was cited as the second strategy, since it improves the environment due to the capacity of trees to retain soil nutrients and water, and prevent severe flooding.

### 6.2.4 Introduction and Adoption of Tree-Based Systems in Claveria

At the turn of the century, 18 % of the municipality was forested by species of *Shorea*, *Pterocarpus*, and other hardwoods. In the 1930s, indigenous people or

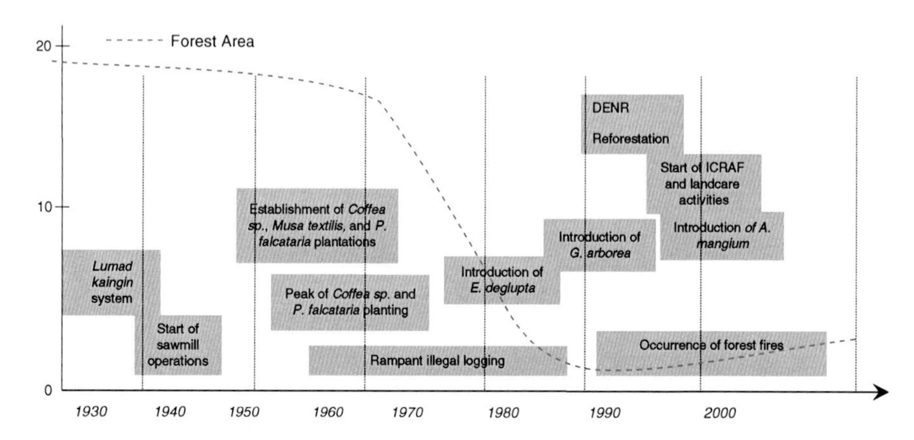


**Fig. 6.3** Change in forest area and forestry-related events in Claveria, Misamis Oriental

*lumads* practiced *kaingin* (slash-and-burn) to clear forest areas and plant annual crops (Fig. 6.3). A marked decrease in the forest area was noted in 1965 due to rampant illegal logging by both large timber concessionaires and small-scale loggers (Garrity and Agustin 1995).

Trees domesticated by farmers included coffee (*Coffea robusta*, *C. excelsa*, *C. Arabica*), gmelina (*Gmelina arborea*), falcata (*Paraserianthes falcataria*), mangium (*Acacia mangium*), mahogany (*Swietenia macrophylla*) and bagras (*Eucalyptus deglupta*). In the 1950s, coffee growing under falcata stands was popular in the area. Falcata was favored by farmers because it provides good shading for coffee, wood for box construction, and pulp for paper.

The Department of Environment and Natural Resources (DENR) introduced gmelina in the 1980s to encourage tree growing as part of its reforestation program. With the establishment of the International Center for Research in Agroforestry (ICRAF) in 1993, agroforestry practices were introduced to farmers. ICRAF initiated the planting of mangium and bagras as agroforestry species.

### 6.2.5 Innovative Agroforestry Practices in Claveria

Tree-based systems commonly used among the respondents were parkland, block planting, border, and hedgerow (Fig. 6.4). These systems could be found in all *barangays* and elevation classes. Originally starting from a simple combination of trees in crop areas, the Claveria farmers developed innovations in the agroforestry systems they practiced.

Out of the 300 households surveyed, 72 % adopted agroforestry in their farms. The parkland system was the most widely practiced (30 %), followed distantly by the hedgerow system or natural vegetative strips (NVS) (18 %), and block planting (16 %). Only 8 % of the respondents adopted border planting.

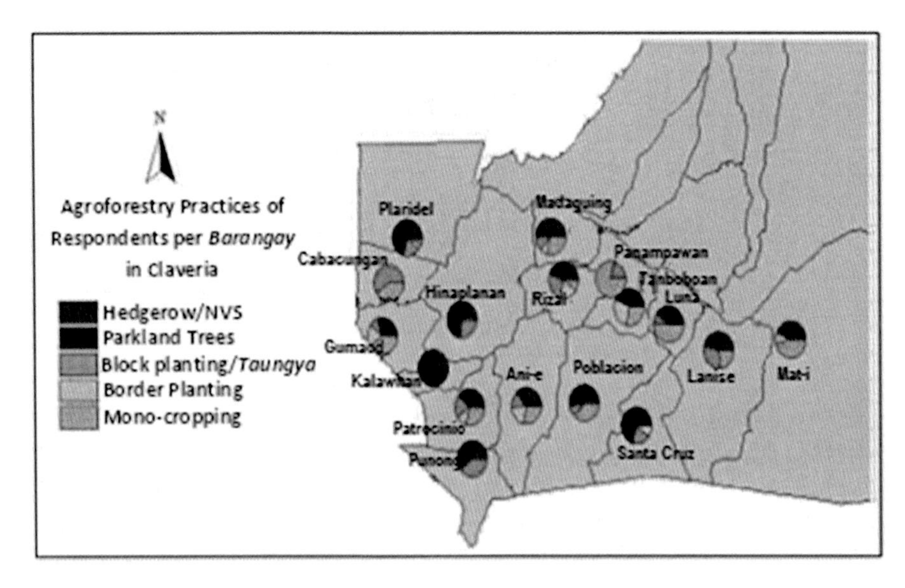


Fig. 6.4 Map showing agroforestry practices of respondents in the study site

**Table 6.1** Reasons or motivations of farmers in adopting certain agroforestry systems in Claveria, Misamis Oriental, Philippines

| Reasons | Block planting (%) | Parkland (%) | NVS/hedgerows (%) | Border planting (%) |
|---|---|---|---|---|
| Income | 48 | 37 | 21 | 27 |
| Soil conservation | 17 | 13 | 55 | 17 |
| Fruit/diversity of harvest | 6 | – | 14 | 14 |
| Use of rolling lands | 4 | 12 | – | – |
| Nature restoration | – | 8 | – | – |
| Construction materials | 9 | 5 | – | 10 |
| Optimal use of land | – | – | 4 | – |
| Wind break | – | – | – | 8 |
| Trend-following/curiosity | – | – | 4 | – |

### 6.2.6 *Farmers' Motivations for Planting Trees*

In all the agroforestry systems, the common reasons given by farmers for planting trees were for additional income, soil conservation, future procurement of construction materials for their houses and furniture, harvest of fruits for home consumption and sale, and to maximize use of their rolling lands (Table 6.1).

## 6.3 Tree Growth, Crop Productivity, and Water and Nutrient Flows in *Gmelina arborea–Zea mays* Hedgerow Systems in Claveria, Misamis Oriental

Miole et al. (2011) evaluated the nutrient flows in *G. arborea–Zea mays* agroforestry systems in Claveria. Experimental plots (18 m upslope × 10 m across) with narrow (1 × 3 m) and wide (1 × 9 m) hedgerow spacing were established. Maize was planted in the alley areas in between hedgerows planted with *Gmelina* trees. Soil profile was characterized following standard procedure for soil profile description. Bulk soil samples were collected and analyzed for routine soil analysis. Tree height and diameter at breast height were measured, while maize plant height, biomass, and nutrient content were determined. Monthly tree leaf litterfalls were collected using 1 $m^2$ litter traps. Incident precipitation, stemflow, and throughfall were measured at each rain event. The collected rain water was analyzed for $NO_3^-$, P, and K contents. A soil erosion plot (6 × 4 m) was established within each experimental plot for sediment and surface run-off measurements. Run-off volume was determined using a water meter installed in a plastic container. Nutrient contents of sediment load and run-off water were analyzed following standard methods.

### *6.3.1 Water Dynamics in Agroforestry Systems*

Stemflow, throughfall, crop biomass residues, and litterfall were classified as processes contributing to nutrient inflows, while soil erosion, surface run-off, and crop harvest represented processes for nutrient outflows. Tree leaf litterfall was the major source of N input in both hedgerow spacing treatments. Maize crop residue was another source of N input for wide hedgerow spacing. The major avenue for N loss in narrow hedgerow spacing was soil erosion. In wide hedgerow spacing, N was lost mainly via crop harvest.

About 54 % of the incoming rainfall landed in the hedgerow system as throughfall, and only 1 % as stemflow, in both spacing treatments. Depths of stemflow and throughfall were slightly higher in the narrow spacing treatment, while depth of surface run-off was higher in the wide spacing treatment. Water passing through stems and barks of trees had higher K content than throughfall. Throughfall, however, showed higher $NH_4^+$ and $NO_3^-$ contents.

### *6.3.2 Growth Performance of Trees and Maize*

The growth performance of trees along the hedgerows and maize crop in the alley areas was better in wide hedgerow spacing than in narrow spacing. Nutrient contents of maize biomass and *Gmelina* leaf litterfall, particularly N, were likewise higher in the former. Providing wider spacing for hedgerow trees would give greater opportunity for a crop to grow well, and consequently, higher maize biomass and yield, with correspondingly higher levels of NPK nutrients.

## 6.4 Enhancing the Food Security of Upland Farming Households Through Agroforestry in Claveria, Misamis Oriental, Philippines

A combination of techniques was used to gather information on the food security situation of farmers in Claveria, including a household survey, focus group discussions, field experiments, bioeconomic modeling using the Water, Nutrient, and Light Capture in Agroforestry Systems (WaNuLCAS) model, and investment and profitability analysis (Magcale-Macandog et al. 2010).

An investment analysis was undertaken to determine the financial viability of investing in agroforestry systems over a 25-year period given a 10 % discount rate. A sensitivity analysis was also performed to compare the economic benefits derived from agroforestry and annual cropping given varying costs of capital ranging from 5 to 30 %.

### *6.4.1 Agroforestry and Improved Access to Food*

The adoption of the agroforestry system has improved food supply and helped address food insecurity among the farmers in Claveria. Fruit-bearing trees combined with crops increase the households' access to food. In any combination, whether planted along the borders, on hedgerows or blocks, or scattered on the farm, they provide additional harvest for marketing, as well as food for farm families.

Usually scattered or planted along borders of farms, fruit trees such as jackfruit, coconut, marang, mango, and avocado serve as additional sources of nutrients or alternative sources of food during the lean months of supply (Brown 2003). While waiting to harvest their main crops, farming households rely on fruits for their nutrition.

### *6.4.2 Increasing Food Access Through Augmentation of Income*

The farmers in Claveria are primarily driven to agroforestry because of the additional income it can provide (Table 6.1). Agroforestry increases income levels and builds assets that improve purchasing power (Garrity 2004). The profitability analysis showed that agroforestry systems were more efficient in utilizing scarce resources and provided higher returns on farmers' investments than annual cropping. Among the different agroforestry systems, the most profitable was the hedgerow system with a net present value (NPV) of PHP 100,817 ha and annualized net income of PHP 11,107 ha. This was an increase of 137 % compared to the income of farmers practicing annual cropping.

The second most profitable agroforestry system was block planting with NPV of PHP 97,013 ha. Under this system, farmers with only one parcel of land would usually allocate 50–65 % of the total farm area to trees and 35 % to crops. Farmers optimized the use of the land for food crop production by growing annual crops under the timber trees during the first three years of tree growth when sunlight could still penetrate the tree canopy (Dalmacio and Visco 2000). The least profitable among the agroforestry systems was parkland planting, with NPV of PHP 60,216 ha. This could be attributed to the small number of trees planted on the farm. Only about 20–25 % of the area could be allocated for timber and fruit trees that were dispersed around the farm, hence the main crops contributed the greater portion of the total income.

It is worth noting that, while timber trees provided income for the households, about 23–56 % of the total benefits from agroforestry were obtained from fruit trees. When corn was not available for consumption and for market, farmers sold bananas and coconuts to generate income with which to purchase rice and other foods. Hence, farmers preferred fruit trees to timber trees.

## 6.5 Predicting the Long-Term Productivity, Economic Feasibility, and Sustainability of the Smallholder Hedgerow Agroforestry System Using the WaNuLCAS Model

This study aimed to predict and assess the long-term productivity, economic feasibility, and sustainability of a *Eucalyptus*-maize hedgerow agroforestry system using the WaNuLCAS model. The results of simulation here were compared with those obtained from simulation of a continuous maize mono-cropping system (Magcale-Macandog and Abucay 2012).

### *6.5.1 Bioeconomic Modeling Using WaNuLCAS*

The WaNuLCAS model was used to determine the complementary effects of trees and crops in the agroforestry systems to improve crop yields. The parameterization of the model used primary data from field experiments as well as secondary data.

WaNuLCAS is a process-based model of the water, nutrient, and light capture in agroforestry systems developed by van Noordwijk et al. (2004). It enables evaluation of the choice of tree species, their spacing and pattern, and possible intercrop.

Parameter values for the model were gathered through field measurements, literature survey, and interviews with farmers. The parameters were used as inputs into the WaNuLCAS model, either directly or indirectly, through sub-routines including Functional Branch Analysis (FBA), Pedotransfer, and WOFOST.

#### 6.5.1.1 Tree and Crop Database

A survey form to describe trees (Treeparam) was designed together with the WaNuLCAS model for parameterization of the tree component (van Noordwijk and Mulia 2002). Default values for the various parameters of the maize crop were available in the crop database of the model. The tree and crop parameters in the databases were part of the input parameters required in the simulation using the WaNuLCAS model.

#### 6.5.1.2 WaNuLCAS Parameterization

The land-use scenarios modeled were the continuous maize (*Z. mays*) cropping system and the bagras (*E. deglupta*) and maize hedgerow agroforestry system. A seven year-old bagras hedgerow with a spacing of $1 \times 10$ m intercropped with maize was selected for the simulation. N and P fertilizers were applied at rates of 15.65 and 16.65 g/m$^2$, respectively, for all plots. P was applied at planting, while N was applied 30 days after emergence. Climatic data were collected from the field site and the Misamis Oriental State College of Agriculture and Technology (MOSCAT) Agromet station. Soil samples were collected from the field and analyzed for chemical and physical properties.

### *6.5.2 Water Balance*

In both the *Eucalyptus*-maize system and maize mono-cropping, subsurface flow and water draining into the different zones and layers accounted for more than half of the water balance in the systems. Water loss through soil evaporation was higher in maize mono-cropping. However, canopy evaporation was higher in the hedgerow system than in the maize system. Water uptake by maize was higher than by the trees in the *Eucalyptus*-maize hedgerow system.

### *6.5.3 Soil Loss*

Continuous maize cropping resulted in greater cumulative soil loss (>100 ton/ha) than the *Eucalyptus*-maize hedgerow system (60 ton/ha) (Fig. 6.5). The greater

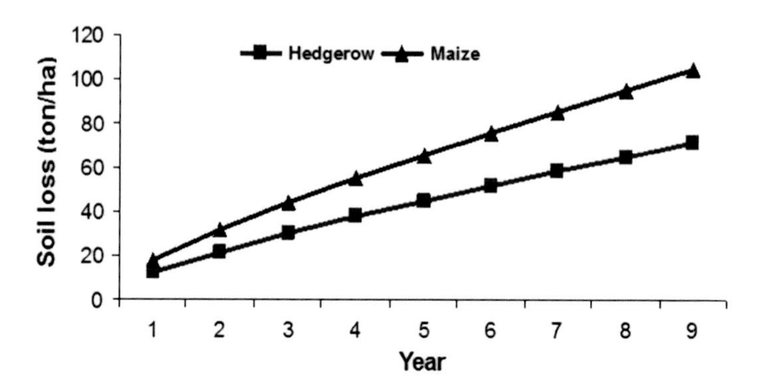


**Fig. 6.5** Simulated cumulative soil loss in *E. deglupta*-maize hedgerow and maize mono-cropping systems

cumulative soil loss in the former was due to cultivation of soil for raising maize. On the other hand, *E. deglupta* hedgerows served as barriers to minimize loss of eroded soil from the system.

### 6.5.4 Crop Yield and Biomass

Simulation results using the WaNuLCAS model showed that maize yield was initially higher in the continuous annual cropping system (2.4 ton/ha) than in the *Eucalyptus*-maize hedgerow system (1.5 ton/ha). This indicates significant competition for light between trees and crops under the latter, as reflected in the low crop yield (Fig. 6.6).

However, yields of maize under both systems declined through the years. Yield from the mono-cropping system exhibited steeper decline than in the *Eucalyptus-maize* hedgerow system. This could be attributed to the greater loss of rich topsoil due to erosion in maize mono-cropping. Tree biomass, on the other hand, increased as the trees grew during the simulation period (Fig. 6.6).

### 6.5.5 Private Benefits of the Two Land Use Systems

The benefits obtained were grain yield from the maize mono-cropping, and grain yield and timber from the *Eucalyptus*-maize hedgerow system. Cost benefit analysis showed the *Eucalyptus*-maize hedgerow system having higher NPV after 9 years of simulation than the continuous maize (PHP 304,323 vs. PHP 20,872).

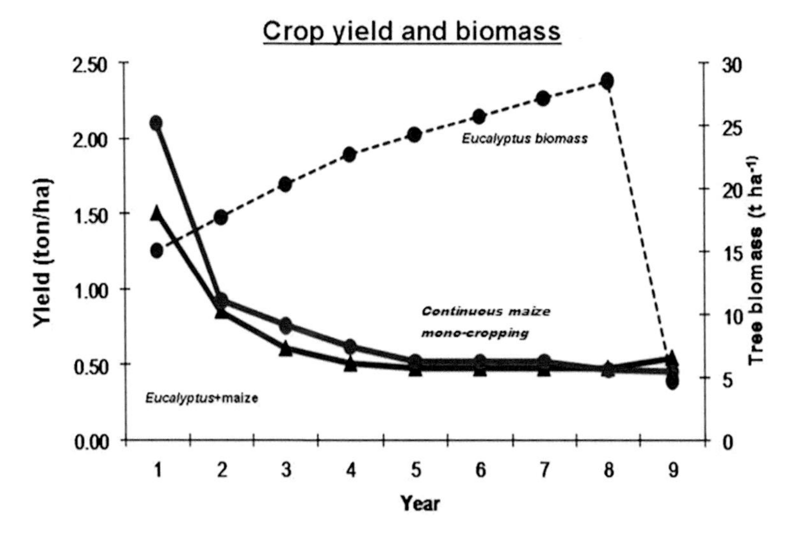


**Fig. 6.6** Simulation results using the WaNuLCAS model showing crop yield under the *Eucalyptus* hedgerow system and annual mono-cropping system as well as *Eucalyptus* biomass in Claveria, Misamis Oriental, Philippines, after running for a 9-year continuous cropping period

## 6.6 Conclusion

Farmers in Claveria develop innovations in the type of agroforestry systems they adopt. The agroforestry system chosen by farmers varies with their motivations. If the primary motive is to obtain additional income, they adopt block planting, border planting, or a parkland system. For soil conservation, farmers prefer the hedgerow system or natural vegetative strips.

The results of the present study have shown that tree spacing affects the nutrient and water dynamics of hedgerow agroforestry systems. Wide hedgerow spacing had higher nutrient inflows and outflows while narrow hedgerow spacing resulted in higher depths of stemflow and throughfall.

The introduction and adoption of an agroforestry system among upland farmers has enhanced their earning capacity and food security. Agroforestry is an essential part of the effort to feed the hungry people in the uplands. While agroforestry efforts cannot substantially alter the social, economic, and political factors that cause food supply inequalities, they can help build up household food security. The integration of trees in agroforestry systems can also help prevent land degradation, increase biodiversity, and at the same time allow the continued use of the land for agricultural crop production.

**Acknowledgment** This study is part of the Smallholder Agroforestry Options for Degraded Soils (SAFODS(ICA4-CT-2001-10092)) project funded by the European Union. We would like to thank our farmer respondents for their time, as well as the valuable experiences and knowledge they shared with the research team. The survey enumerators who helped us in gathering information and data are also duly acknowledged.

## References

Brown DM (2003) Considering the role of landscape, farming system and the farmer in the adoption of trees in Claveria, Misamis Oriental Province, Philippines. MSc Sustainable Agriculture and Rural Development, Imperial College, University of London

Bureau of Soils (1985) Detailed reconnaissance soil survey suitability classification: Claveria complementation project, Claveria, Misamis Oriental. Bureau of Soils, Manila

Dalmacio RV, Visco RG (2000) Agroforestry concepts, principles, and practices: training manual on agroforestry for sustainable development and people empowerment. College of Forestry and Natural Resources, University of the Philippines Los Baños, College, Laguna

Domingo EV, Buenaseda MGB (2000) Land and soil resource accounting: the Philippine experience. In: Proceedings of the international workshop on environmental and economic accounting, Manila, 18–22 September 2000

Escaño CR, Tababa SP (1998) Fruit production and management of slopelands in the Philippines. Food and Fertilizer Technology Center. http://www.agnet.org. Accessed 8 Feb 2009

Garcia-Barrios L, Ong CK (2004) Ecological interactions, management lessons and design tools in tropical agroforestry systems. Agroforest Syst 61:221–236, In: Nair PKR, Rao MR, Buck LE (eds), New vistas in agroforestry: a compendium for the 1st World Congress of Agroforestry. Kluwer Academic, Netherlands

Garrity DP (2004) Agroforestry and the achievement of the millennium development goals. Agroforest Syst 61:5–17

Garrity DP, Agustin PC (1995) Historical land use evolution in a tropical acid upland agroecosystem. Agr Ecosyst Environ 53:83–95

Israel DC, Briones RM (2012) Impacts of natural disasters on agriculture, food security, and natural resources and environment in the Philippines. Discussion series 2012-36. Philippine Institute for Development Studies

Leaky R (1996) Definition of agroforestry revisited. Agroforest Today 8(1):5–7, Nairobi: ICRAF

Magbanua RD, Garrity DP (1990) Agroecosystems analysis of key upland farming systems research site. In: Proceedings of the 1988 acid upland design workshop. International Rice Research Institute, Los Baños

Magcale-Macandog DB, Abucay ER (2012) Predicting long-term productivity, economic feasibility and sustainability of smallholder hedgerow agroforestry system using the WaNuLCAS model. Ecosyst Dev 3(1):51–58

Magcale-Macandog DB, Visco RG, Delgado MEM (2006) Agroforestry adoption, innovations and smallholder farmers' motivations in tropical uplands of Southern Philippines. J Sustain Agr 28(1):131–143

Magcale-Macandog DB, Rañola FM, Rañola RF, Ani PAB, Vidal NB (2010) Enhancing the food security of upland farming households through agroforestry in Claveria, Misamis Oriental, Philippines. Agroforest Syst 79(3):327–342. doi:10.1007/s10457-009-9267-1

Miole RN, Visco RG, Magcale-Macandog DB, Abucay ER, Gascon AF, Castillo ASA (2011) Growth performance, crop productivity, and water and nutrient flows in *Gmelina arborea* Roxb.-*Zea mays* hedgerow systems in Southern Philippines. Philippine J Crop Sci 36(3):34–44

Paningbatan E (1989) Soil erosion problem and control in the Philippines. In: Decena F Economic values of erosion in upland farms in Rizal and Batangas, Philippines, 1989–1994. Master of Science Thesis in agricultural economics, University of the Philippines Los Baños College, Laguna

Sajise PE, Ganapin DJ Jr. (1991) An overview of upland development in the Philippines. In: Blair G, Lefroy R, Technologies for sustainable agriculture on marginal uplands in Southeast Asia. Proceedings of a seminar held at Ternate, Cavite, 10–14 December 1990, pp 31–44

van Noordwijk M, Mulia R (2002) Functional branch analysis as tool for fractal scaling above- and belowground trees for their additive and non-additive properties. Ecol Model 149:41–51

van Noordwijk M, Lusiana B, Khasanah N (2004) WaNuLCAS version 3.1, Background on a model of water nutrient and light capture in agroforestry systems. International Centre for Research in Agroforestry (ICRAF), Bogor

Villancio VT, Lapitan RL, Cabahug RD, Arboleda LP, de Luna CC, Paelmo RF, Papag AT, Solatre JS (2003) Sustaining agriculture and forestry through agroforestry initiatives of people in the upland: cases in the Philippines

Wise R, Cacho O (2002) Tree-crop interactions and their environmental economic implications in the presence of carbon-sequestration payments. Working Paper CC11, ACIAR Project ASEM 2002/066, http://www.une.edu.au/feb1/Economic/carbon/

# Chapter 7
# Managing Environmental Risks and Promoting Sustainability: Conservation of Forest Resources in Madagascar

**Bruno Ramamonjisoa**

**Abstract** From 1990 to 2005 Madagascar implemented a national environmental policy comprising three successive programs. However, after 15 years of implementation, the results were not as initially expected. Forest degradation had continued unabated, resulting in significant environmental problems, such as loss of soil fertility, siltation of rice paddies and water bodies, and further reduction of water supplies in cities. As a result, food availability in rural areas had deteriorated, aggravating the already high levels of poverty. Sustainable forest management had therefore become crucial to sustainable living, yet the problem was how to implement it in the context of increasing poverty. Assuming that the forest could be sustainable only if it contributed to increasing the income of local people, a method of non-extractive valorization of forests was developed and tested. The method incorporated a range of activities that highlighted the ecosystem services of forests, including environmental education, ecotourism, no tillage agriculture, and the development of rural markets for non-wood products. These activities were expected to incentivize the local community to change their behavior and preserve forest and ecosystem services. However, the approach remained highly localized because it attracted immigration from other areas. This case study therefore concluded that sustainable forest management could not be achieved without integrated management at the regional and national levels, as well as on a local scale. Future research will need to focus on understanding migration, on successful spatial management methods, and on identifying alternative economic incentives for stakeholders.

**Keywords** Forest management • Local community • Local scale • Sustainable living

---

B. Ramamonjisoa (✉)
Water and Forest Department, School of Agronomy,
University of Antananarivo, Antananarivo, Madagascar
e-mail: bruno.ramamonjisoa@gmail.com

N. Kaneko et al. (eds.), *Sustainable Living with Environmental Risks*,
DOI 10.1007/978-4-431-54804-1_7, 

## 7.1 Introduction

Madagascar is the world's fourth largest island, situated in one of the most biodiverse regions on earth. However, large-scale environmental degradation has occurred on the island. From the 1980s onwards, the International Monetary Fund and the World Bank advocated structural adjustment programs for most African countries, and these led to the establishment of environmental programs and the formulation of strategies for poverty reduction (World Bank 2002). Implementation of such programs and strategies has implicitly become a criterion of eligibility for debt reduction programs. Thus, in addition to a structural adjustment program including market liberalization and the privatization of state companies, in 1989 Madagascar adopted a National Environmental Action Plan (NEAP). It was based on a program spread over 15 years with the primary objective of conserving natural resources, and was monitored to ensure sustainable economic development and improved quality of life. Since the implementation of the NEAP and economic policies linked to foreign direct investment, several additional policies and strategies have also been implemented. Yet, despite more than 15 years of environmental programs and nearly USD400 million in financing, Madagascar's problems have never been solved. Indeed, in 1989, policymakers found that the relevant statistical data were inaccurate because of the general lack of tools to precisely assess the situation and the scale of its evolution. To this day Madagascar still has no tools to measure indicators of environmental quality.

Agricultural production remains the dominant economic activity, providing one third of GDP and 80 % of foreign exchange earnings. However, because of slash and burn practices, persistent fires, and the resulting reduction in forest cover to protect the vast watershed of Madagascar, erosion is still present in almost all areas. This leads to lower soil fertility, increased costs associated with water and road infrastructure, and damage to marine ecosystems. Sustainable management of natural resources is therefore essential to avoid adverse environmental consequences, including loss of species, landslides, drying up of springs, and silting up of lowlands. Forests play a crucial role in stabilizing the agricultural system as well as the soil, and make an important contribution to reducing environmental risk.

Despite the implementation of an environmental program in 1990 and a forest policy in 1997, the forests have continuously deteriorated. They are subject to the combined effects of skimming, which removes many of the species of commercial interest, and slash and burn, which completes the extraction of woody vegetation considered to be of no commercial value (Fig. 7.1). These two causes of forest degradation have never been eradicated because no reform has ever directly addressed the two practices in question. There were proposals to introduce the concept of *KoloAla* (sustainable production forests) to convert the logging industry's mode of operation to a tender model (based on inventories of existing timber volume on forestry plots). However, no significant changes have been made so far. Meanwhile, the same slash and burn practices continue on the basis of a government ordinance dating back to 1960 (Ramamonjisoa 2004).

**Fig. 7.1** Slash and burn in Madagascar's eastern region

There is now better understanding of the reasons why forest resources continued to deteriorate despite the implementation of the NEAP and its USD400 million funding. Sustainable forest management has become key to sustainable living, yet the problem remains how to implement it in the context of increasing poverty. It is assumed that the forest can be sustainable only if it contributes to increasing the income of local people.

## 7.2 Theoretical Explanation

A sustainable forest management system will be successful only if the community is involved in all activities. Community-based forest management is a promising strategy for achieving this aim that has been applied in Africa over the past two decades. It has the potential to enable mitigation of forest resource degradation throughout the South, and particularly in Africa.

Community-based management will flourish only if it has a sustainable financing system that can redistribute profits to all stakeholders. Based on this assumption, data coming from Madagascan experiences between 2006 and 2010 were used to simulate how business as usual could be changed into standardized forest use (Ramamonjisoa 2010, 2012). The results of this simulation showed that strengthening community management by reorganizing forest use and redistributing its benefits was beneficial both for the forest users themselves (i.e., loggers and households) and for the local area (leading to construction of public infrastructure, for example). It also conferred decision-making power at the local level (Bertrand and Montagne 2008; Ramamonjisoa 2010a, b). The only way to compensate for loss of income from slash and burn, fuel wood production, and other extractive activities, therefore, is to improve methods for non-extractive forest valorization that benefit the local population. Such activities have to be distributed equally among households.

## 7.3 Case Study

### *7.3.1 Context*

From 1990 to 2005 Madagascar implemented a national environmental policy comprising three successive programs. After 15 years of implementation the results were not as initially expected, and forest degradation had continued unabated. Natural forest cover (not including secondary forests) was 10.7 million ha in 1990 (18.1 % of the national territory), 9.7 million ha in 2000 (16.5 %), and 9.2 million ha in 2005 (15.7 %) (Ministry of Environment and Forest 2007). The forest degradation had also resulted in significant environmental problems, such as loss of soil fertility, siltation of rice paddies and water bodies (Fig. 7.2), and further reduction

**Fig. 7.2** Flooding of rice fields in Madagascar's central region

**Fig. 7.3** Soil erosion and road degradation in Anjozorobe (north east of Antananarivo)

of water supplies in cities. As a result, food availability in rural areas had deteriorated, aggravating the already high levels of poverty. Sustainable forest management had therefore become crucial to sustainable living.

The poor who have to fund their daily subsistence tend to exploit their natural resources—forests, fisheries, and mines. Land that is ecologically fragile and unsuitable for agriculture is subjected to intensive farming and overgrazing. This causes soil erosion (Fig. 7.3), depletion of water resources, lower yields for crops, and desertification. Women and children in poor households often spend several hours a day fetching water and firewood—time and energy that would be better spent on schooling,

**Table 7.1** Formal and informal governance

| | Formal governance | Informal governance |
|---|---|---|
| Annual funds generated | USD20 million | USD73 million |
| Actors | State | State |
| | Donors | Forest service (departments) |
| | Technical partners | Primary banks |
| | International NGOs | Shipping companies |
| | National NGOs | Traders (collectors-intermediates-importers-exporters, etc.) |
| | Civil societies | |
| | Engineering offices | Customs services |
| | Decentralized services (including forestry) | Decentralized communities (*fokontany*,[a] municipalities, regions, etc.) |
| | Communities (farmers) | Communities (labor collection and/or sampling) |

[a]The smallest subdivision of local government

community development, and building good family relations. Land degradation and poverty in rural areas encourage mass migration to already crowded cities, where the lack of clean water, sanitation networks, and garbage collection services contribute to water pollution and disease. Poverty, being associated with high rates of infant mortality, motivates households to have additional children to ensure that more potential breadwinners will survive.

### *7.3.2 Poor Governance and Inequality in Income Distribution*

An analysis of Madagascan forest governance highlighted the competition between:

- Formal governance (imposed by donors and conservation NGOs), focused on protection and conservation by means of protected areas, and
- Informal governance (dominated by private operators), based on the exploitation of mining and non-wood forest products (Ramamonjisoa 2005)

The state expects to receive an annual income of nearly USD28 million, which prompted it to divide the entire national territory, including protected areas, into mining squares. This caused juxtaposition of natural resource use (use of the same resource for different economic activities).

Meanwhile, the trade in forest products has still not stopped. In fact, it has actually increased in volume, making environmental management policies focused on conservation even more difficult to implement. The trade in wildlife species has been estimated at a minimum of USD500,000 per year, while the turnover of wood products amounts to around USD73 million per year.

As ever, the degradation of forest resources remains due to the fact that the two types of governance (formal and informal) are both based on monetary incentives

(Table 7.1). Formal governance cannot replace informal governance because the funds received through formal governance represent only one-quarter of the revenue from illicit practices. In addition, the funds are not well distributed among the actors involved.

The system of conditionality imposed by the structural adjustment programs and misinterpretation of the term "foreign direct investment" have helped to establish priorities that are not necessarily based on the economic viability of natural resources. In addition, the establishment of many non-state institutional structures has helped to generate an antagonism between the administration and conservation NGOs.

Degradation is also explained by the existence of gaps and inconsistencies in legal norms, contributing to shortcomings in management system implementation. In the absence of a forest control system (due to lack of means) these gaps tend to generate and sustain informal norms that are based not on the effect of coercive laws, but on the monetary incentives generated by the marketing of products obtained illicitly.

The juxtaposition of resource use also creates a conflict in the application of legal norms, as well as an inefficient system of coercion of users by the state (Ramamonjisoa 2004). By prescribing two distinct uses for the same land, the state creates inefficiencies in the management system, as in the case of mining squares in protected areas (Fig. 7.4). As a result, procedures are no longer effective and the actors have to negotiate for the administrative services required, such as logging permits and transportation authorization. The service providers therefore become the players who make the decisions, but the factors informing such decisions are economic and social, rather than technical. Consequently, the administration becomes easily corruptible.[1] In addition, institutional inconsistencies prevent administrators from managing natural resources effectively, while decentralized services become permeable to the intrusion of related issues and purely political. The final explanatory factor in poor governance, moreover, is the behavior of citizens, who tend to regard the necessary procedural documents as products for sale. Almost all such documents, including business licenses, office notices, and instructions, have become commodities. Forest agents say that their equipment and budgets are insufficient, and the lack of performance-based pay demotivates them from working for the public interest, driving them to consider themselves as at the service of particular interests.

---

[1] The determinants of corruption in the forestry sector are many and varied: (1) the remoteness and isolation of forest workers, (2) inadequate checks and sanctions for several years, (3) insufficient work resulting in dependent agents becoming operators, (4) hierarchical political pressure exerted through a network of complicity leading to malfunctions in the procedures and records, and (5) complexity and lack of clarity of records, combined with inadequate updating.

Significant financial and human behavior-related determinants have also been noted, including:

- Inadequate remuneration for officials and the scarcity or absence of compensation in case of displacement;
- Financial avidity among operators;
- Lack of ethics and practice models.

(JURECO Studies and Consulting 2009)

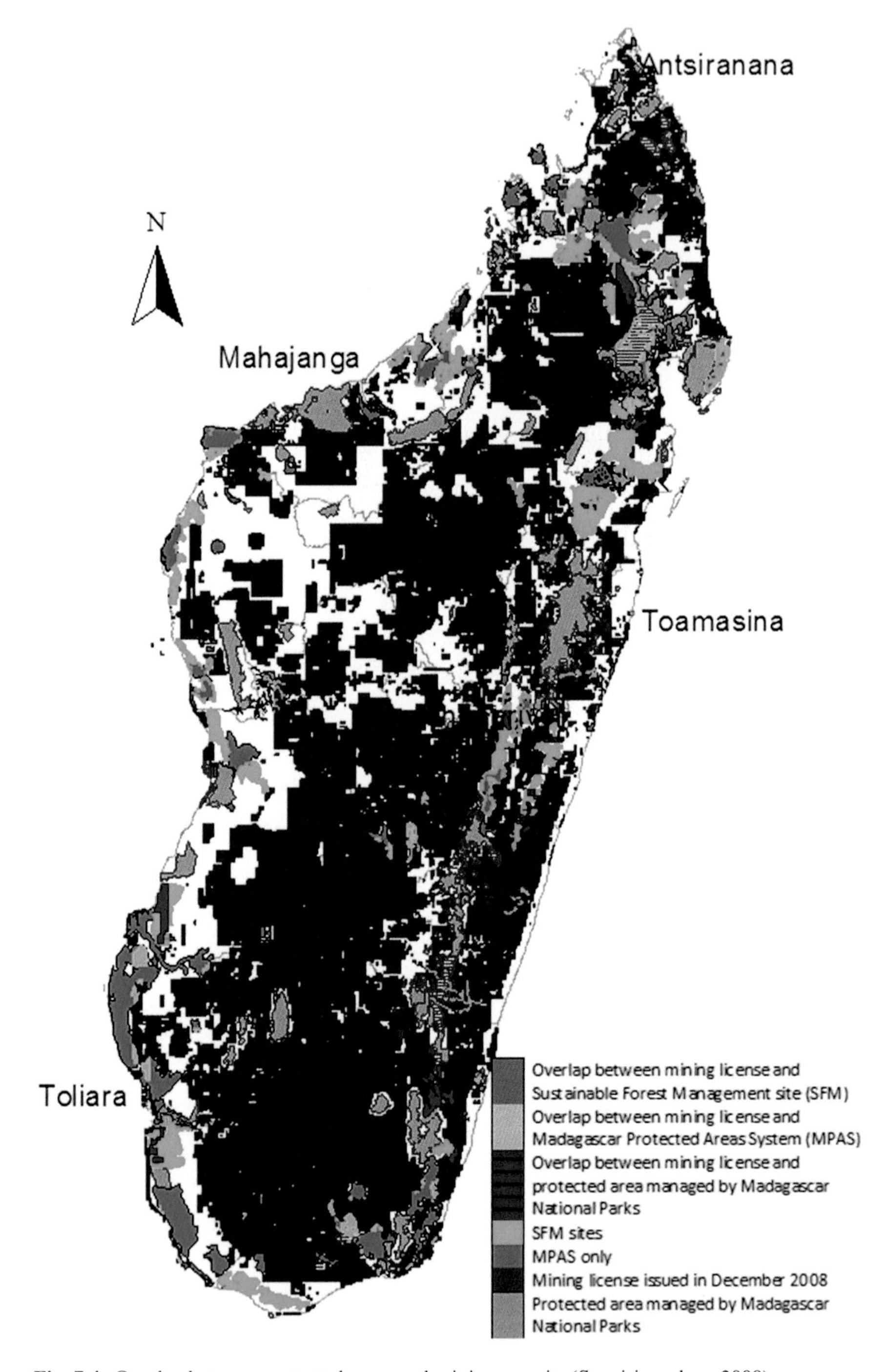


**Fig. 7.4** Overlap between protected areas and mining permits (Soanirinavalona 2008)

**Table 7.2** Sources of power and associated rationales

| Rationale | Economic | Social |
|---|---|---|
| Source of power | – Possession of a skill or a functional specialization difficult to replace (the expert is the only one to have the know-how, knowledge and experience)<br>– Control of relations with the environment<br>– Communication | Social position based on:<br>– Prestige (external experience/money)<br>– Nobility (inheritance)<br>– Anteriority (age) |
| Type of power | Authoritative: characterized by a form of power requiring obedience accepted without coercion, as well as trust and respect | Compassionate: a mixture of persuasion and manipulation that seeks to evoke emotion, and to compel someone to do what he did not really want to do |

*Source*: Ramamonjisoa 2005

Two solutions were proposed to solve the problem: the establishment of a redistributive tax, and capacity building. The proposed introduction of a redistributive tax was based on the principle of focusing action on how we can better capitalize on the existing informal organization, rather than trying to change it. Redistributive taxation should be a tool that works both as an incentive (for state actors in general) and as a means of enforcement (for other players who do not follow the new norms to be established). However, the big issue with this new means of paying for environmental services is the fact that introducing the tax would require the establishment of a new independent organization to manage forest resources. On the other hand, the formulation and implementation of redistributive taxation is expected to solve the budget problems of decentralized and deconcentrated state services (including the forest service, the customs service, decentralized authorities, and municipalities). The forest service, for instance, could make more than twice its entire annual budget by taking 10 % of the total income (amounting to approximately USD73 million per year) from the illegal trade of timber products.

Meanwhile, capacity building is another prerequisite for changing behavior by means of public policies in the short and medium term. In addition, education over the long term can enable the switch to a non-hierarchical coordination system supporting the implementation of sustainable governance to maintain effective forest management.

The inability of formal governance to replace informal governance reflects failures in the development and implementation of public policies. These failures are in fact explained as follows: Sources of power (Table 7.2) within societies are usually based on the possession of a skill or a functional specialization that is difficult to replace, as well as control of environmental relations and communication (Crozier and Friedberg 1996). In Madagascan society, however, power is conferred by prestige, nobility (based on inheritance), and social position based on anteriority (age) in a dominant-dominated hierarchy that is accepted by all and determines access to and use of Madagascar's natural resources (Ramamonjisoa 2005).

The evidence indicates that informal governance has led to a number of norms based on a dominant-dominated hierarchical coordination system. Ever since the initial implementation of structural adjustment, this has resulted in prestige (based on external experience/possession of money) replacing competence as the criterion when choosing agents to take charge of natural resource management. Within this context, the actors developed strategies to capture funds at many levels. However, networks are better at capturing funds (because they have a greater redistributive capacity), and it is therefore critical that all stakeholders act in terms of institutional strengthening focused on organization and standards.

The implementation of redistributive taxation was expected not only to help establish a management system adapted to the constraints. In addition, it was expected to generate significant financial resources that could contribute to regional and even national development. No doubt the system would improve the availability of cash at the local and regional levels. In practice, however, trials conducted in the north west of Madagascar (the GESFORCOM Project[2]) met with only qualified success.

Community management of forests cannot withstand transactions of external origin that do not lead to local financial autonomy: In Madagascar the USD1.25 million injected for the implementation of GELOSE[3] (between 1994 and 2004) neither increased the income of communities, nor contributed to financial volume in the local zones where management transfers were negotiated and contracts signed (Ramamonjisoa and Rabemananjara 2012). In this case, the emergence of local elites overlooking the community undermined efforts. In fact, decentralization provides numerous examples of how village councils can serve to strengthen the local elite, rather than providing a community voice. During transfers of forest management, stakeholders reinterpret "negotiated" standards, taking advantage of the relevant laws. Therefore, despite the transfer, the public administration may continue to play a role in legalizing forest degradation practices, albeit implicitly, and taking premiums from those practices for which the legal basis of the transfer system is unknown or incomprehensible.

Standardized forest use without economic incentives cannot succeed if it is not applied throughout the country or the region. The viability of the system depends on illicit practices ceasing to exist. Moreover, if the share of benefits offered by the new system remains insignificant, illegal logging will not be prevented. The conclusion, therefore, is that projects must not be isolated, but should be rolled out across the country or region to stop the existing leakage of funds. This would require a new forest policy, a new action plan, and, above all, control of the forests.

In summary, poor governance and inequality in distribution of income have become key issues because of stakeholder behavior and the ever-increasing forest degradation it induces.

---

[2] Community and municipality forest management (Gestion forestière communale et communautaire) was a project funded by the European Union between 2007 and 2010 in three countries (Madagascar, Mali, and Niger). The goal of the project was to comparatively test improved forest management methods in the three countries. Activities were focused on the development of forest management plans, the reinforcement of control systems, and the development of new taxation systems, the proceeds of which were shared among the main managers of forest resources, such as the forest service, the local community, and the municipality. See http://www.gesforcom.eu

[3] Gestion locale sécurisée (secured local management)

### 7.3.3 *More Income, Less Logging*

Development of a range of activities that highlight the ecosystem services of forests—including environmental education, ecotourism, no tillage agriculture, and the development of rural markets for non-wood products—is the only alternative solution to address all the constraints, including poor governance, inequality in the distribution of income, and poverty. Tested in the national park of Ranomafana and elsewhere, this method demonstrated that income flows into rural areas directly as a result of certain activities such as environmental education, ecotourism, and the development of rural markets for non-wood products.

Entrance fees for these parks represented only 30 % of a tourist's total travel expenses, while the other 70 % were indirect expenses paid directly to local stakeholders (in return for accommodation, food, or handicrafts) (Ramamonjisoa 2000). The increased income for local communities resulted in a 0.07 % decrease in forest degradation in these areas (Ministry of Environment and Forest 2007).

A trial is being conducted in the Mandraka station to try to overcome the gaps in the case study above. The Mandraka station is one of the four forest training stations of the School of Agronomy's Water and Forest Department, situated 70 km east of Antananarivo. The site includes fragments of rainy natural forests and areas reforested using native species. Ecotourism incorporating an environmental education system was set up within the local community, as well as a no tillage agriculture system and a rural market for non-wood forest products. Local communities needed to be involved soon after the activities were launched to ensure the success of the projects (Figs. 7.5 and 7.6).

## 7.4 Key Issues and Recommendations

Unfortunately, the trial described above is not certain to halt forest degradation because it induces additional constraints such as migration from the poorest regions in the south to these areas, while it does not successfully inhibit slash and burn practices. By general consensus, direct degradation is attributable to the spread of slash and burn for rice or maize cultivation, taking place even inside protected areas.[4] Both migration and slash and burn are regarded as key drivers of deforestation and will be tackled to ensure that a sustainable management system can be implemented. A system of no tillage agriculture could replace slash and burn practices,

[4] A study conducted in 2011 focused on assessing the current threats and biological impact caused by traffic of precious woods and hunting of lemurs, concluding that: (1) deforestation accelerated at a rate of 0.36 % per annum between 2005 and 2010 in the Masoala National Park, and 0.71 % per annum in the Marojejy National Park, and (2) analysis of satellite photos clearly showed some glades that always appeared at the edges of the forests. This was not a result of selective logging, but rather of slash and burn activities and other land uses. Glades were almost always connected to the hydrographic network, and were particularly remarkable for the river Ampanio in the Masoala park (LRA 2011).

**Fig. 7.5** Community working—Men in action

while setting up a rural market for non-wood products could prevent forest logging. The strategy is to avoid implementing only one conservation-related activity, but to combine two or three depending on the magnitude of forest degradation. These solutions could be instituted in different regions—including migration departure points, as well as final destinations—and on various scales, and should be regarded as integrated management schemes on a national, regional, or local level.

Some key questions remain to be addressed in the future: How can slash and burn practices be effectively replaced considering the socioeconomic, cultural, and technical aspects of this degradation-inducing agricultural system? Can we expect these practices to be replaced, as the great majority of them are illegal? How can rural markets for non-wood products that really benefit local farmers be established? In addition, research needs to focus on understanding migration, on successful spatial management methods, and on identifying alternative economic incentives for stakeholders.

Full comprehension of these phenomena could help in establishing realistic strategies to minimize environmental risk and promote sustainability. This is the only way to fight scarcity and thus break the vicious cycle of poverty and deforestation.

**Fig. 7.6** Community in action—A gender-based approach



## References

Bertrand A, Montagne P (2008) Domanialité, fiscalité et contrôle : la gouvernance locale contractuelle des ressources renouvelables dans un contexte de décentralisation (Niger, Mali et Madagascar). Mondes en développement 141:11–28. doi:10.3917/med.141.0011

Crozier M, Friedberg E (1996) Organization theory. Centre for the Sociology of Organizations, Paris, p 86

JURECO Studies and Consulting (2009) The situation of corruption in the environment & forestry. JURECO Studies and Consulting, Antananarivo, p 77

Laboratoire de Recherche Appliquée (LRA), ESSA-forêts (May 2011) Réalisation d'état des lieux des valeurs universelles exceptionnelles dans les secteurs perturbés des parcs Masoala et Marojejy (capitalisation des acquis), Rapport préliminaire, p 36

Ministry of Environment and Forest (2007) General direction of forest: forest degradation analysis. Conservation International, Antananarivo, p 102

Ramamonjisoa BS (2000) Socio economic context of forest conservation in Madagascar. Conservation International, Antananarivo, p 59

Ramamonjisoa BS (2004) Origin and impact of national policies for natural resource management: the case of Madagascar. J Forestier Suisse 2004:467–475

Ramamonjisoa BS (2005) La reconstruction du système de régulation de l'usage des ressources forestières à Madagascar : la nécessite d'une éducation économique. Mémoire d'Habilitation à Diriger des Recherches. Water and Forest Department, School of Agronomy, University of Antananarivo, p 260

Ramamonjisoa BS (2010) Rapport thématique économie et gestion des filières. Projet GESFORCOM, November 2010, p 52

Ramamonjisoa BS (2012) Governance and political nature in Madagascar Geopolitics and environment the lessons of the Malagasy experience. IRD Collection Objectif Sud, Montpellier, pp 187–206

Ramamonjisoa BS, Rabemananjara Z (2012) Une évaluation économique de la foresterie communautaire. Dans Enjeux et moyens d'une foresterie paysanne contractualisée. Expériences de système de gestion locale à Madagascar. Presse universitaire de Bordeaux Les Cahiers d'Outre-Mer 65(257):125–155

Soanirinavalona RO (2008) Contribution à la résolution des problèmes de cohabitation entre création d'aires protégées et activités minières. Mémoire de DEA en Sciences agronomiques. Département des Eaux et Forets. ESSAgronomiques. Université d'Antananarivo, p 115

World Bank (2002) Annual report 2002, Volume 1, Year in Review, Antananarivo, p 171

# Chapter 8
# Community-Based Mangrove Forest Management in Thailand: Key Lesson Learned for Environmental Risk Management

**Surin On-prom**

**Abstract** This article discusses community-based mangrove forest management and its implications for environmental risk management. The article draws on the case study of Pred Nai village in Trat province, near the Cambodian border in southeast Thailand. The village of Pred Nai has successfully re-forested 1,920 ha of mangrove forests that were previously converted into shrimp aquaculture ponds in the mid-1980s. The village has set up a community forest committee and a community resource use regulation and management plan in order to regulate, control, and manage the use of resources by community members. In the process of community mangrove forest management, local villagers have been encouraged to participate in every single step of forest management and planning. It is the active involvement of the local people together with the support of the relevant authorities and national institutions that makes the Pred Nai example a success. The article concludes by pointing out the key lessons learned from community-based mangrove forest initiatives in Pred Nai that can be applied in natural disaster risk management processes.

**Keywords** Community empowerment • Community participation • Community-based mangrove management

## 8.1 Introduction

In developing countries it is widely recognized that community participation helps to ensure sustainability, makes development activities more effective, and builds local capacity. The participation of local people also ensures equitable benefits for the diverse interest groups within a population as well as ensures effective

S. On-prom (✉)
Faculty of Forestry, Kasetsart University, Bangkok, Thailand
e-mail: fforsro@ku.ac.th

N. Kaneko et al. (eds.), *Sustainable Living with Environmental Risks*,
DOI 10.1007/978-4-431-54804-1_8, © The Author(s) 2014 

stewardship. This is partly because communities often have better knowledge and expertise in the management of local resources than government agencies/private industry. Indeed, the involvement of local people in such development business may help reduce government costs. As such, local participation and decentralization are currently being promoted in many developing countries as an alternative approach to development and resource governance. Brown et al. (2002) argue that such a community-based approach often leads to more equitable and sustainable natural resource management. These authors provided a number of reasons to support their argument. One of those reasons concerns the issue of proximity to resources. Those in closest contact with, and whose livelihoods are impacted by, natural resources are best placed to ensure effective stewardship.

Community-based natural resource management (CBNRM) is an approach under which communities become responsible for managing natural resources (forests, land, water, biodiversity) within a designated area. CBNRM gives communities full or partial control over decisions regarding natural resources, such as water, forests, pastures, communal lands, protected areas, and fisheries. The extent of CBNRM control can range from community consultations to joint management or to full responsibility for decision making and benefit collection, using tools such as joint management plans, community management plans, stakeholder consultations and workshops, and communal land tenure rights. Together with decentralization reforms, CBNRM ensures stakeholder participation, increases sustainability, and provides a forum for conflict resolution (World Bank 2006).

In Thailand, participation of local communities in natural resource management has been recognized and promoted by government agencies. Evidence of this is provided by the current National Constitution, which contains some articles giving rights to local people and communities to participate in such state-owned development projects. These local communities currently have a say in decision making processes related to natural resource management. In the forestry sector, the Royal Forest Department (RFD) promotes community forestry projects where local communities living in or adjacent to national reserve forests are given usufruct rights to resources within the community forest areas. The RFD also provides them with technical and financial support. It is worth noting here that the most economically and ecologically successful CBNRM is achieved when community participation is accompanied by the devolution of ownership rights over resources. Once a community has been given tenure over the resources they are helping to manage, they are able to gain benefits from the use of these resources and their interest in participation is naturally increased.

The village of Pred Nai in Thailand's Trat province has successfully re-forested 1,920 ha of mangrove forest that were previously converted into shrimp aquaculture ponds in the mid-1980s. Community conservation of mangrove forests in Pred Nai village emerged without outsider intervention more than 20 years ago. Clearly, the initiative by the Pred Nai villagers qualifies as community-based mangrove forest management and numerous lessons can be drawn from it. The successful implementation of community forestry has reaffirmed the role and significance of a community-based approach in natural resource management. This community-based approach to mangrove management has achieved the twin objectives of restoring coastal and marine biodiversity, and generating income for all socio-economic groups of the village. The initiative has also been able to integrate the marginalized and poor households in the entire process, and tangible benefits are even generated for inhabitants from surrounding villages (Silori et al. 2009).

The aim of this paper is to illustrate the successful case of Pred Nai community-based mangrove forest restoration and management, and to draw key lessons that can be applied in environmental risk management. Following the introduction, the paper presents the theoretical background to the concept of CBNRM and its implications for sustainable development. It then sets out the case study of Pred Nai village. In conclusion, the paper draws some lessons learned and makes recommendations, as well as offers further research questions.

## 8.2 CBNRM and Sustainability: Theoretical Background

Community-based natural resource management (CBNRM) means different things to different people. The term "CBNRM" not only takes a variety of forms depending on location, or socio-political and bio-physical context, but the term itself is used and interpreted in many different ways (Abensperg-Traun et al. 2011). According to these Abensperg-Traun et al. (2011), the term CBNRM simply describes the management of resources such as land, forests, wildlife, and water by collective, local institutions for local benefits. Following this explanation, the term CBNRM is often associated with programs that (1) are focused on terrestrial wildlife and (2) involve some kind of commercial use of that wildlife in order to generate income for local people.

The World Bank refers to CBNRM as a third alternative to command and control and market-based approaches to natural resource management (World Bank 2006). The term itself has breadth, adaptability, and robustness since it has been described as, *inter alia*, a tool or a set of tools, a checklist, a method, a set of activities, a model, a process, and an approach (World Bank 2006). At the same time, CBNRM in a particular context is considered to be a problem solving mechanism in land and natural resources. CBNRM is relevant to, and has the potential to provide solutions to, some of the problems found within communal lands.

The importance of CBNRM as an approach has recently been realized. There are at least two reasons explaining what is important about CBNRM (WWF 2006). The first reason concerns the willingness of the state to manage natural resources. In newly independent Southern African countries, for example, the governments did not consider the management of natural resources to be a priority. Financial resources were diverted to other sectors (health, education, and infrastructure development). Military-style methods of enforcing natural resource legislation are also no longer considered appropriate. And the second reason concerns the linkages between natural resources and local livelihoods. Most of the populations in developing countries depend on natural resources for their livelihoods. In Southern Africa, for example, most of the population lives in the communal lands. In all communal lands, the harvesting of natural resources plays a significant role in people's livelihoods. This means that strategies and policies to ensure the sustainable management of land and natural resources are important to national economic development. Because of the direct link between people's welfare and the environment, it is important that new approaches and ideas are developed and implemented in natural resource management. The alternative is that people will get poorer and the environmental resources upon which they depend will become scarcer, thereby worsening the already serious poverty trap (WWF 2006).

In addition, CBNRM has been recognized for a long time as a means of promoting safety. CBNRM plays a significant role in promoting a culture of safety by reducing local vulnerabilities and building capacity. Active participation and involvement of communities at the grassroots can make a real difference. This is because through the involvement of local people in the whole process of natural resource management, their feelings and real needs are considered as well as inherent resources. The capacities of local people are enhanced to help them assess situations, identify needs and problems, implement activities, and evaluate outcomes. It can be said that the CBNRM approach emphasizes the involvement of communities, which are at the heart of decision making and implementation of natural resource activities and management.

The links between CBNRM and the issue of resilience were observed and studied. The existing literature on resilience in social-ecological systems strongly suggests that community-based institutions may play a key role in fostering the resilience of communities and the ecosystems they inhabit. Some suggest that local management institutions enhance resilience. One reason is because management practices are locally adapted and based on local ecological knowledge (WWF 2006). More importantly, however, some CBNRM organizations promote social learning, an intentional process of collective self-reflection through interaction and dialogue among diverse participants. Social learning is promoted in part due to CBNRM groups' attention to monitoring and adaptive management, and their emphasis on learning and education. As rural communities face increasing environmental stresses as well as unpredictable economic and political shocks, the ability to learn and adapt is critical to their sustainability and resilience (WWF 2006).

It can be concluded that, through CBNRM, the community is not only the beneficiary but also the main actor in every step of the forest resource management process. The involvement of the community is important to ensure sustainability. The information collected will be more relevant and will reflect the opinions and realities of community members, particularly the vulnerable and poor. The capacity (self confidence, knowledge, skills such as team work, planning, etc.) of the entire community to deal with different situations they face will be developed. Natural forest resource management and community development activities and programs will achieve better, more practical and effective results. Community life will become more stable and sustainable.

## 8.3 Pred Nai Community-Based Mangrove Forest Management

This section presents the case study of Pred Nai villagers and their efforts to conserve mangrove forests. This community-based organization was founded in the mid-1980s to reverse the effects of destructive mangrove logging near the coastal village of Pred Nai. A parallel goal was the recovery of local crab populations, which are an important source of income for poorer members of the village. The success of Pred Nai villagers in restoring the mangrove ecosystem made the group into a model and point of reference for policy reforms that aim to transfer authority to forest communities.

The village of Pred Nai is located in Trat Province on the eastern seaboard of Thailand. The geographical area of the village is 378.7 ha, of which the inhabited area covers nearly 42 %, while the remaining area is under agriculture and other land use practices. The area of community-managed mangrove forest, one of the last surviving mangrove forests in Thailand's eastern seaboard, is spread over approximately 12,000 rai (1,920 ha), about 1 km to the west of the village. The mangrove forest is regularly inundated by 12 major and 6 minor creeks (Fig. 8.1).

Pred Nai village was founded in the 1850s by about 8–10 households. In 2009 it was reported that the population of Pred Nai village had increased to 560 people or nearly 130 households. The majority of the households own land, though most of

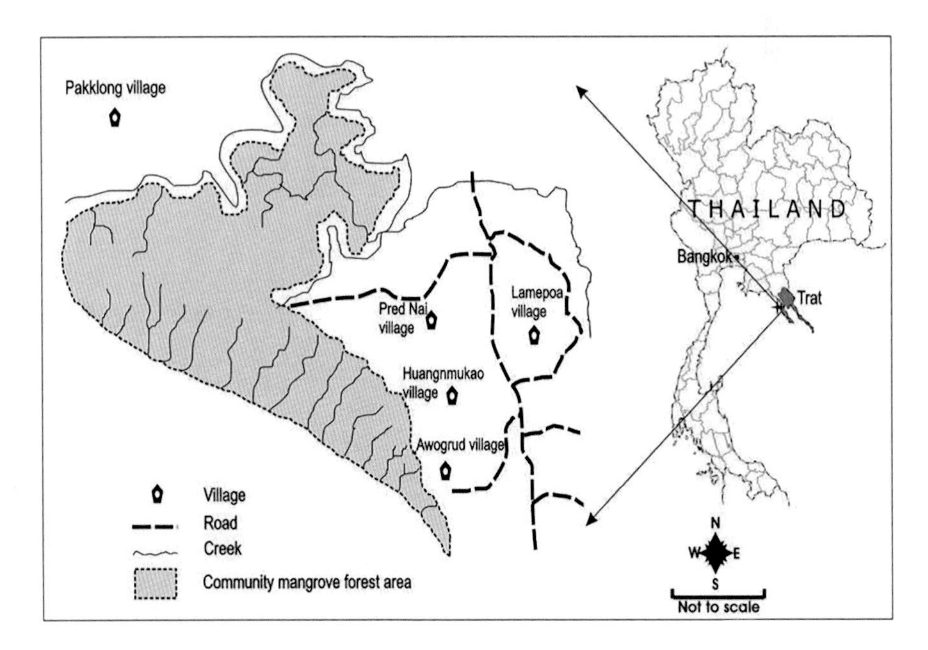


**Fig. 8.1** Pred Nai village map (*Source*: Silori et al. 2009)

them have marginal to small holdings. Only around 7 % of the households do not own land. The agriculture land use is dominated by rubber plantations and orchards of durian, rambutan, jackfruit, mango, and mangosteen. Fish and shrimp farming is another major land-use practice in the village. Small and marginal land holders also work for daily wages in the agricultural sector, while a sizeable number of households venture into the mangroves to collect various marine animals for home consumption and for sale such as grapsoid crab (*Metopographus* sp.), mud crab (*Scylla serra*), and shrimps, as well as a variety of fish and shells.

Agriculture and daily wage labor in the agricultural sector, fish and shrimp farming, and collection of marine products from mangroves were reported as major income sources by the villagers. Silori et al. (2009) claimed that the average annual income of Pred Nai villagers ranged from about THB 65,000 (USD 1,914) for landless households to slightly above THB 1million (USD 30,531) for a large land holder, averaging about THB 435,000 (USD 12,809) per annum. Fish and shrimp farming contributed up to 63 % to the average annual income, followed by agriculture (26 %), and collection of marine animals (mainly grapsoid crab) (9 %). The remainder was made up of other activities such as collection of honey and agro-tourism. Thus, nearly 74 % of the total average household income can be directly or indirectly attributed to mangrove and marine resources (Silori et al. 2009).

The mangroves of Pred Nai were placed under a logging concession in 1941. Uncontrolled logging and intensive shrimp farming caused heavy destruction. As a result, an area of nearly 48,000 ha of mangroves was reduced to 1,920 ha by the early 1980s (Silori et al. 2009). Government concessions favored corporations and restricted the villagers from harvesting crabs, shellfish, and other mangrove resources. Some local people converted degraded mangrove areas into shrimp farms and built gates to block seawater. This activity further damaged the mangrove ecosystems and availability of crabs and other marine animals decreased. Other illegal activities such as timber harvesting and hunting of birds and other wild animals further degraded the biodiversity of the mangroves. Rapid depletion of mangroves and marine resources directly threatened the livelihoods of the local villagers and they therefore decided to stop the degradation process. A 'core group' of 5–10 villagers started to resist the corporate destruction of the mangroves, and their resistance rapidly gained momentum and support from the rest of the villagers, intensifying demands to stop the logging. The struggle of the villagers paid off and commercial logging was stopped in 1987 by the Government. The logging companies were ousted from the area and the sea water gates were destroyed to allow the flooding of the degraded mangroves. This was an important incentive for the villagers to institutionalize the struggle of the core group and they formed a village level group called Pred Nai Community Forestry Group (PNCFG) in 1987. In due course, the PNCFG was supported by local and provincial governments and also by donor and technical agencies. The Regional Community Forestry Training Centre for Asia and the Pacific (RECOFTC), for example, was one of the main technical support organizations, and assisted in the formulation of a mangrove restoration and management plan following a participatory approach.

Since its emergence the PNCFG has conducted various activities to regenerate the mangrove forests. The group initially developed a forest management plan that produced intensive mapping of forest resources and instituted forest patrols to prevent illegal logging and charcoal production. Also, the group initiated tree-planting activities in degraded mangrove areas. Over time, some stands have begun to regenerate naturally. Since 1993, the Pred Nai initiative has focused on the sustainable management of marine resources and ecological monitoring within the mangrove forests. In 1997, the group introduced some regulations on crab harvesting. These rules specified the suspension of harvesting during the crab breeding period in October (UNDP 2012).

In parallel with the conservation activities, the PNCFG also initiated a village community fund with the help of a local monk, primarily with the objective of introducing the concept of financial sustainability of community initiatives. Under this initiative, the villagers were encouraged to save part of their income and earn some interest, while at the same time keeping the savings within the village so that the funds could be loaned to other needy households. Later, this initiative won the support of government agencies and NGOs, obtaining donations from international agencies and local institutions such as the Social Investment Fund (SIF), the Thailand Research Fund, the Education Institute, and even from political parties. The village common fund has total savings of more than THB 3 million at present, and regularly receives contributions from crab catchers, local traders, and also from agro-tourism activities, and thus it is sustained year after year (Silori et al. 2009).

Over nearly three decades the capacity of the PNCFG has been strengthened through various training programs provided by many organizations. RECOFTC has been one of the key organizations among these. It has been associated with the PNCFG since 1999 and helped to promote the PNCFG's outreach to communities both within and outside of Thailand. It has also encouraged the exchange of ideas about sustainable utilization of natural resources among the villagers. As a result, the PNCFG has taken the lead in building a network of more than 20 village-level community forestry groups along the four provinces at the eastern seaboard of Thailand to manage mangrove resources. Not only this, but Pred Nai has also received international attention for its contributions in the community forestry sector. Besides attracting a number of study tours from many countries, the PNCFG received the Green Globe Award in 2001 and was nominated for the Equator Initiative award in 2004 (Senyk 2005). According to the villagers, the capacity of the PNCFG to manage the resources, regulate their access and use, and resolve conflicts has increased. At the same time, the relationship with the key government departments, such as the Department of Fisheries, the Tambon (Sub district) Administrative Office (TAO), and the Department of Marine and Coastal Resources (DMCR) has improved. Villagers interact regularly during monthly village meetings. Awareness of the importance of mangrove forests and their conservation has been raised. Also, key government officials are involved in formal and informal village meetings and capacity building activities, and help in conflict negotiations, if any.

There can be no doubt that the conservation efforts of the Pred Nai villagers and the PNCFG have had a direct impact on restoring biodiversity, alleviating poverty, and facilitating local economic development. The chief biodiversity impact has been the restoration and protection of 1,920 ha of coastal mangrove forest (UNDP 2012). This substantial regeneration of mangrove forest cover has enabled the return of wildlife species to the coastal area. It was reported that stocks of crab, shellfish, and fish had all increased (Silori et al. 2009). The restoration and conservation of mangrove forests also improves the long-term sustainability of the villagers' economic activities. The household income patterns clearly reflect the importance of coastal and marine resources in the livelihood strategies of the local residents, especially the poor ones. This is further substantiated by the trends in the number of villagers who venture into the mangroves to collect marine animals, particularly mud crabs. About a decade ago, 6–7 people used to venture into the mangroves every night to collect crabs. This number at present has grown to nearly 70. More important, however, is the fact that, despite such a steady increase in the number of crab collectors, the average quantity of crabs per visit for each collector has not declined over time. It stands at around 7–8 kg, while the time spent has declined significantly to an average of 4–5 h per night at present, as compared to almost a whole night 10 years ago (Silori et al. 2009; Somsak et al. 2004).

In conclusion, the steady growth of conservation and development activities in Pred Nai since the mid-1980s is testament to the activities' resilience and the broad-based support they receive from villagers and various development agencies. Meanwhile, the substantial management capacity developed through the PNCFG ensures that the initiative is well-equipped to sustain its impact over time.

## 8.4 Discussion

In this final section, an attempt will be made to summarize the lessons learned from the Pred Nai case study that may be relevant to and apply in the sustainability of disaster risk management. As is widely recognized, disaster management is a pressing issue for all of us and should be undertaken on a comprehensive basis. Such an approach to natural disaster risk management should engage, involve, and empower communities so that community members can cope with the adverse effects of natural hazards. And the most effective approach for achieving sustainability in dealing with uncertainties and risks at a local level is the establishment of resilient communities. The major lessons drawn from Pred Nai village are as follows.

The case of Pred Nai suggests that sustainability of development programs (mangrove rehabilitation and conservation in this case) by a community largely depends on the benefits that can be generated and the local people's participation. If they understand and recognize the significance of the mangrove forest and the benefits that can be generated from activities related to it, the community will step in and take part. This case also made us realize that the community-based management process was based on local people's participation, initiated by a group of villagers who started to provide others with relevant information about the destruction of the mangrove forest and marine life. These people instigated popular awareness of the responsibility of everyone in the community—not just certain people—to rely on themselves collectively, without the need to wait for an agency to come in to provide assistance. If the community does not act to solve a problem right away, such problems eventually impact livelihoods. If a community can mobilize its partners, however, they can contribute to achieving the same goals of protecting the resources and preventing them from being destroyed. Often the necessary participation can be achieved through the representative systems of the group or the committee that has been mandated to deal with or to make decisions on such issues as quickly as possible.

A community-based approach is a critical factor in sustainable development at community and national levels. The most common elements of a community-based approach are participation, partnership, empowerment, and ownership by local people. The emphasis of sustainable natural resource management should focus on local communities and the people who live in or adjacent to them. Unless the conservation and development efforts are sustainable at individual and community level, it is difficult to achieve the goal of sustainability. However, there need to be opportunities for the poor and "marginalized groups" in the village to be involved from the initial programming stage of conservation and development activities. In the case of Pred Nai, for example, the poor were allowed to participate in the process of formulating harvesting regulations. Through the community-based activities, these marginalized groups should be able to participate alongside community organizations like PNCFG, government officials, experts, and NGO workers as the key stakeholders.

Partnership development and institutional linkages on all horizontal and vertical levels are critical to achieve the sustainability of conservation and development activities. As we observed from Pred Nai, the PNCFG has incorporated innovative partnerships and a wide range of participants. At local level, the group has networked with several villages to enable dialogue with those stakeholders in coastal and marine resources and also to ensure broad support for sustainable harvesting in the mangrove forests. Moreover, the group has gained experience in working collaboratively with outsiders such as RECOFTC, fishery experts, and foresters. Relationships with outside institutions and organizations have also been important for helping the group to overcome funding and technical challenges to implement projects. The provincial governor became an active supporter of Pred Nai's initiatives. More importantly, the group has been working with Tambol Administrative Organization (TAO), the elected local government body, in order to formalize regulations on coastal and marine resource harvesting.

It is worth making the generalization here that community-based approaches and empowerment of communities help to achieve sustainability in conservation and development. However, CBNRM efforts need technical, financial, and institutional support. Institutionalizing the community can result in more sustainable development programs.

## 8.5 Conclusion

In order to ensure more efficient CBNRM, this paper recommends: (1) legalizing the status of community-based organizations like PNCFG in order to enhance their capacity to make decisions concerning natural resource uses and allocations;

(2) providing necessary support for Local Government Organizations LGOs to ensure that they can identify effective and sustainable mangrove forest management strategies through the process of local people's participation; (3) compiling and managing knowledge, experience, and innovation emerging from community mangrove forest management for use in education, learning, and dissemination to other interested communities and individuals; and (4) developing an effective monitoring and evaluation system for community mangrove forest management.



## References

Abensperg-Traun M, Roe D, O'Criodain C (eds) (2011) CITES and CBNRM. In: Proceedings of an international symposium on "The relevance of CBNRM to the conservation and sustainable use of CITES-listed species in exporting countries," Vienna, Austria, 18–20 May 2011. Gland, Switzerland: IUCN and London: IIED

Brown D, Malla Y, Schreckenberg K, Springate-Baginski O (2002) Community forestry, from supervising 'subjects' to supporting 'citizens': recent developments in community forestry in Asia and Africa. Natural Resource Perspectives 75. Overseas Development Institute (ODI), London

Senyk J (2005) Lessons from the Equator Initiative: community based management by Pred Nai Community Forestry Group in the mangroves of Southeastern Thailand. http://www.umanitoba.ca/institutes/natural_resources/pdf/Tech%20Report%20Thailand%2 0-%20Jason%20 Senyk.pdf

Silori CS, Soontornwong S, Roongwong A, Enters T (2009) Links between biodiversity conservation and livelihood security in practice: a case study of community conservation of mangroves in Pred Nai, Thailand. Final draft manuscript for Community Forestry International workshop, September 15–18, 2009, RECOFTC, Pokhara

Somsak S, Worrapornpan S, Kaewmahanin J (2004) Biodiversity management and poverty alleviation by Pred Nai community forest, Trad province, Thailand. Ferng-fah Printing Company Limited, Bangkok

United Nations Development Programme (2012) Pred Nai mangrove conservation and development group, Thailand. Equator Initiative Case Study Series, New York

World Bank (2006) Agriculture investment sourcebook. http://www.worldbank.org

WWF (2006) The community–based natural resource management manual. Wildlife Management Series. World Wide Fund for Nature, Harare, Zimbabwe

# Part II
# Ecosystems, Food Security, and Disaster

# Chapter 9
# Necessity of Adaptive Risk Management for Fisheries and Wildlife

**Hiroyuki Matsuda**

**Abstract** The conventional theory of ecosystem and population management does not include the concept of risks. In risk management, survival rates and reproductive rates are taken into account in mathematical models describing conditions of the ecosystem, and uncertainty and environmental variation when measuring are taken into account when predicting population sizes. As a result, future prediction can only be made in a probabilistic way. Even when a general prediction is made about the future, therefore, it is rare to predict it accurately. Risk evaluation to show a prediction range with some allowance is required. Moreover, if management measures are predetermined, it is impossible to cope with a contingency. Adaptive management is therefore recommended as a management method to cope with uncertainties. This entails making a management plan based on unverified assumptions, continuously monitoring changes in the situation while implementing the plan, adjusting the management measures as required, and verifying the appropriateness of the assumptions used. It is important to predetermine how the measures will be adjusted and the assumptions verified. In this chapter, we highlight the difference in thinking before and after adaptive risk management is established, describing two cases: management of fisheries resources and wildlife management. Numerical calculations that appear in the figures in this chapter can be obtained through the website (http://risk.kan.ynu.ac.jp/matsuda/2014/SLER.html) and additional tests can be conducted with Excel files. Eager readers are strongly recommended to try this.

**Keywords** Deer management • Environmental impact assessment • Maximum sustainable yield • Uncertainty

---

H. Matsuda (✉)
Graduate School of Environment and Information Sciences, Yokohama National University, Yokohama, Japan
e-mail: matsuda@ynu.ac.jp

N. Kaneko et al. (eds.), *Sustainable Living with Environmental Risks*,
DOI 10.1007/978-4-431-54804-1_9, © The Author(s) 2014

## 9.1 Is It Really Good for Fisheries to Reduce the Fish Population by Half?

Capture fishery is an industry that enables humans to utilize wild seafoods as a resource, although, if overfished, breeding individuals become short of supply, making it impossible to secure the resources in the future. Accordingly, theories of resource management for appropriate fishing have long existed, although such theories typically ignore uncertainty. Of the conventional resource management theories, one such theory that ignores uncertainty is the classical theory of "maximum sustainable yield," or MSY, that appears in fisheries and wildlife management (Clark 1990).

In the classical MSY theory, we consider the reproduction curve as shown in Fig. 9.1. The formulae are shown at the end of the chapter. Biological populations are characterized by exponential growth in population numbers. Calculated in the simplest possible way, the per capita growth rate of a population would increase constantly, like the relationship between the amount of a bank deposit and interest (corresponding to the thin line in Fig. 9.1). However, the per capita growth rate in fact decreases as the stock increases because availability of food, habitat, and other resources becomes limited. Therefore, the surplus production, or the product of the population number and the per capita growth rate, forms a peak, corresponding to the bold curve in Fig. 9.1. In addition, the population does not increase beyond a threshold, which is called carrying capacity, denoted by $K$ in Fig. 9.1. The surplus production produces a curve of one crest in relation to the stock, as in Fig. 9.1, and with an intermediate stock level, the largest catches can be continued. The catches are called MSY, which in the case of Fig. 9.1 is 500,000 t. If the surplus production is larger than the catches (implied by the broken line in Fig. 9.1), the resources increase. If it is smaller, they decrease. The equilibrium stock biomass under the given catch level is the intersection of the surplus production curve and the catch

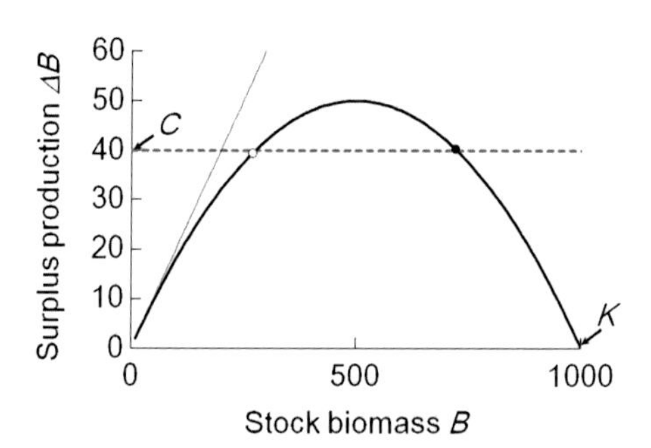


**Fig. 9.1** The conceptual relationship between stock biomass ($B$) and the surplus production (*bold curve*, $\Delta B$). $\Delta B$ is not proportional to $B$, unlike the thin line. $\Delta B$ reaches a peak at $B=500$ and $\Delta B$ becomes 0 at $B=1{,}000$, or carrying capacity $K$. If $\Delta B$ is larger than catch amount (*broken line*, $C$) the stock will increase. The stock will decrease if $\Delta B<C$. When $C<$MSY or 50, there are stable and unstable equilibria that satisfy $\Delta B=C$, as shown by *black* and *white circles*, respectively

amount, indicated by white and black circles in Fig. 9.1. If the yearly catches are smaller than the surplus production, the stock can be maintained at a high level. But if the resources approach the carrying capacity, the surplus production decreases, and this is not effective utilization. If, on the other hand, the catch amount is larger than the MSY, the stock decreases until extinction.

Catch at the MSY seems to be desirable from the viewpoint of sustainability, but fishery operators rarely know the true MSY level. Consequently, fishing is continued at a higher rate than the MSY, thus endangering many fish species.

Indiscriminate fishing damages not only the ecosystem but also economic profits over the long run. But why do attempts to increase production not succeed? The reasons are roughly divided into two types, as has long been pointed out in the science of fishery (Clark 1990); they include economic discounting and the tragedy of the commons. We do not explain these in depth here, however, as a bigger problem is the very assumption that the relationship as shown in Fig. 9.1 exists. The ecosystem is uncertain, non-equilibrium, and complex, yet in Fig. 9.1 these three factors are not considered. They will be explained sequentially below.

First, unless the reproduction relationship as shown in Fig. 9.1 is known, the MSY cannot be achieved. If 600,000 t is continuously caught when the MSY is 500,000 t, the resources will be exhausted.

Next, Fig. 9.1 implies that if catches are continued at a certain level, the stock will settle at a certain amount, but this notion would in fact be laughed off by fishery operators. Populations of marine organisms fluctuate greatly depending on the environment, and while prohibition of fishing may even decrease populations in some cases, in other cases populations may not decrease even if many more individuals are caught. That is to say, the ecosystem is unsteady, and dynamic sustainability must be considered accordingly.This challenge can, however, be overcome by the adaptive management described later. In such cases, the prospects for resource recovery can be shown only in a probabilistic way. That is, risk management is required.

Finally, Fig. 9.1 relates only to the fish species to be caught, but an increased population of a certain fish species concerns not only the stock of that particular species, but also the population sizes of the organisms on which they prey and their natural enemies. The MSY theory does not consider this point, either. The effects of such interaction among species lead to a need to manage the ecosystem as a whole. Theoretically, this differs considerably from population management (Matsuda and Abrams 2004, 2006, 2013).

The southern bluefin tuna (*Thunnus maccoyii*) was intensively caught from the 1960s to the 1980s, resulting in stringent fishery controls being agreed on by an international management organization, the Commission for Conservation of Southern Bluefin Tuna (CCSBT). They set a numerical target to recover the stock to its 1980 level by 2020, and in the latter half of the 1990s signs finally appeared to suggest that the resources would eventually be recovered. Then Japan substantially increased its catches without obtaining the consent of other countries, and was brought to the international court. Yet, after Japan reduced its catches again, southern bluefin tuna resources did not recover as expected. One reason was that the resource recovery prediction at the beginning was too optimistic (Mori et al 2001, see Fig. 9.2)

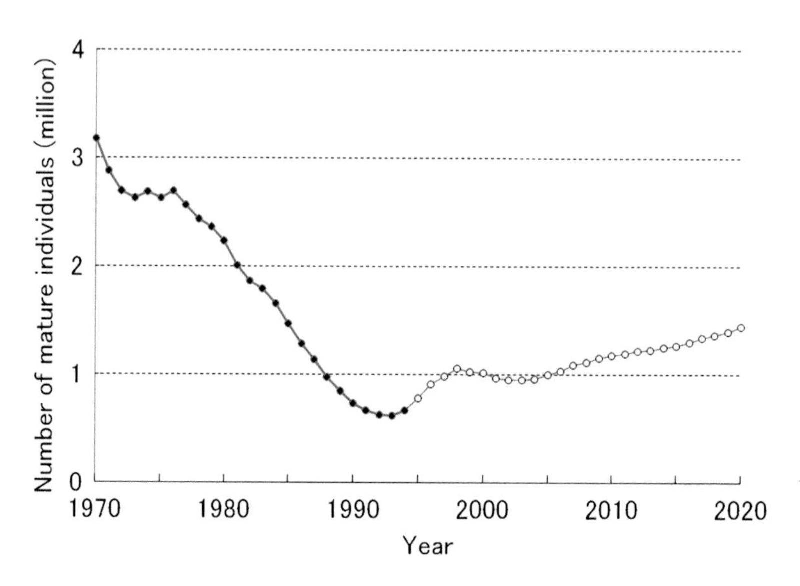


**Fig. 9.2** Trends in the southern bluefin tuna population (Mori et al. 2001). *Black circles* and *white circles* are past estimates and medians of future projections, respectively

and the other reason was under-reporting of catches. Mori et al (2001) predicted that the stock would not monotonically increase even under the stringent fishery controls, which was indeed the case.

## 9.2 Comparison Between Conventional Management Methods and Adaptive Management Considering Uncertainty

To overcome the problems of uncertainty and non-equilibrium, the adaptive management method was proposed (Holling 1966; Walters 1986). This entails making a management plan based on unverified assumptions, continuously monitoring while implementing the plan, adjusting the management measures as the situation evolves, and verifying the assumptions used.

The idea of adaptive management has been introduced in the recent management of marine resources. An example is shown in Fig. 9.3. With the same mathematical model shown in Fig. 9.1, the prediction when catching is continued at the exploitation rate (0.05) to realize the theoretical MSY is as shown in Fig. 9.3a. If, however, the true MSY is not known, the exploitation rate often becomes excessive. Even if the figure shows that the stock changes along a smooth curve, therefore, it is not persuasive because it does not accord with the actual experience of fishery operators. They know that the stock fluctuates greatly year by year because of environmental changes.

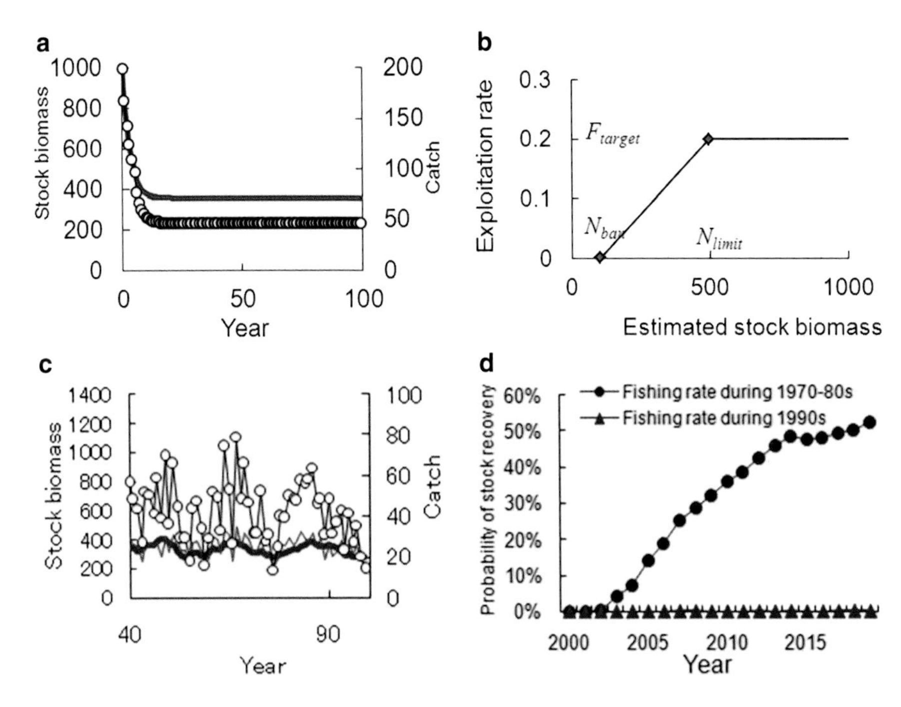


**Fig. 9.3** Numerical examples of adaptive fisheries management using the same mathematical model as Fig. 9.1. (**a**) Trends of stock (*bold line*) and catch (*circles*) under constant recruitment ratio and constant exploitation rate (exploitation rate $F$ is 0.2). (**b**) The rule to determine the exploitation rate ($F$) for the total allowable catch as a function of recent estimate of stock biomass. (**c**) Trends of stock (*bold line*) and catch (*circles*) under the rule given by panel b with estimation errors in stock abundance (estimated stock is shown by *thin line*) and process errors. (**d**) Probability of stock recovery under exploitation rate during 1970–1980s and exploitation rate during 1990s in a Japanese chub mackerel fishery (Kawai et al. 2002)

On the other hand, Fig. 9.3b illustrates the biological allowable catch decision rule, whereby the exploitation rate has been adjusted in accordance with estimated stock values in recent years to determine the Total Allowable Catch (TAC). Here the exploitation rate during a period of high stock is deliberately and mistakenly set at twice that in Fig. 9.3a. Because the true stock value is not known, the possibility of such a mistake cannot be ignored. Even in such a case, selective fishing can be avoided and relatively stable catches can be maintained as shown in Fig. 9.3c because the exploitation rate is lowered as the resources decrease.

Stock trends shown in Fig. 9.3a may be unrealistic because errors in the recruitment process are ignored. The figure shows a deterministic process whereby the stock trends are monotonically increasing or decreasing. Figure 9.3b, c seem more realistic, as these are stochastic processes incorporating process errors. Therefore, as shown in Fig. 9.3d, the vertical axis is the probability of stock recovery (to a particular level) instead of the stock biomass in Japanese stock assessment under various scenarios (Kawai et al. 2002).

In Fig. 9.3, three types of uncertainty are considered. One is *measurement error*; an example is the disagreement between the thick and thin lines in Fig. 9.3c. This occurs because the true stock is unknown. The second is *process error*, and an example is the zigzag changes in stock in Fig. 9.3c, which are caused because the intrinsic rate of natural increase fluctuates every year. This is ignored in Fig. 9.3a. The third is *operational error*. This means that the actual exploitation rate varies slightly from the exploitation rate set in Fig. 9.3b. In risk management, it is necessary to consider any or all of these three types of uncertainty.

In addition, we should recognize the lack of full certainties relating to the mechanism of resource change. This is not considered in the management plan in Fig. 9.3. Uncertainties consist of those not predicted by the manager, those predicted but not clearly considered in the management plan, and those considered in risk management. The latter are uncertainties within the scope of the management plan's assumptions. However, not all uncertainties can be considered.

In Fig. 9.3c, numerical calculations are conducted by taking uncertainty into account and withdrawing random numbers. Every time a random number is withdrawn, therefore, a different result is obtained. In the case of the exploitation rate rules of Fig. 9.3b, of 100 numerical calculations, the minimum stock became lower than 300 t twice, and the minimum catches became lower than 20 t six times during 100 years, with no measures taken to prohibit fishing. An important role of risk management is to set some situations to be avoided, evaluate the probability (risks) of them occurring, and make a management plan to ensure that they will be held within allowable limits.

In conventional marine resources management, general predictions were made ignoring uncertainties such as those shown in Fig. 9.3a, and a management plan to obtain the most desirable situation was devised. Such a plan was unable to cope with uncertainty. The predictions made in such a management plan will be different from actual future results.

Since we used a stochastic model, we cannot describe unique future projections for each management scenario.Therefore, we calculate the risk of management failure from a large number of simulation results as shown in Fig. 9.3d. Note that the vertical axis of Fig. 9.3d is the risk of management failure, instead of the future stock abundance.

## 9.3 Do Not Make a Single Prediction (Japanese Deer Protection Management Plan)

Populations of Japanese deer have increased too much in numerous locations in Japan, posing a social problem. To address this issue, mathematical models of population dynamics were made. For example, the Forest Life Section of the Forestry and Forest Products Research Institute created a simple simulation program to manage deer, as shown in Fig. 9.4. Calculations were made with an Excel program, Simbambi, developed by Dr. Horino, section chief of the institute. The initial Simbambi was designed to make a general prediction for a specific capture policy as shown in

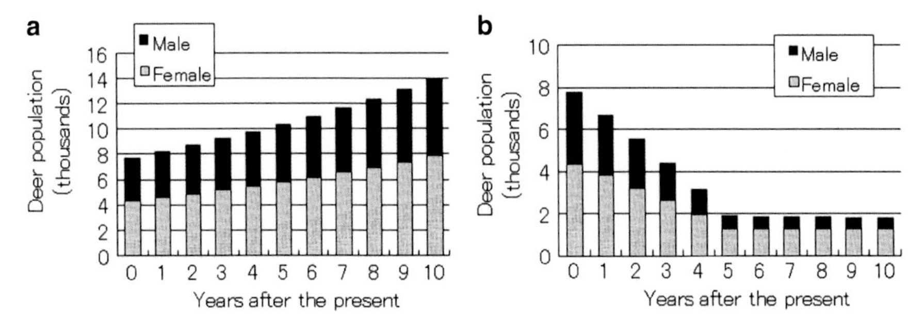


**Fig. 9.4** Projection of deer population of Mt. Goyozan under two different scenarios: (**a**) No capture and (**b**) 800 individuals are caught every year until the 5th year and 190 individuals are caught every year from the 6th year

Fig. 9.4, without considering the aforementioned uncertainty, and was a numerical calculation for the region of Mt. Goyozan, Iwate Prefecture. The simulated population size was subsequently changed from the actual size in order to create the present revised simulation program, Simbambi 4.1. The Japanese deer in Mt. Goyozan are considered excessive, and it is known that they must be captured or they would rapidly increase (Fig. 9.4a). In this model, the population size would increase limitlessly, but in reality it is believed that the food would be completely eaten, resulting in mass death. In fact, mass death is reported in such places as Cape Shiretoko, Nakajima Island in Lake Toya, and Kinkazan. Figure 9.4b shows predicted population size when 400 males and 400 females are captured every year for 5 years. In the region of Mt. Goyozan, the relationship between captured numbers and changes in population size is predicted when implementing the population management plan, the aim being to achieve coexistence between humans and deer.

Actually, however, the plan failed. From 1990, the number of deer captured was increased to 1,000 or more, and from 1993, the number of does captured was increased to about 1,000 more than planned for 5 years. The true population size is unknown, but it was probably underestimated, since it was possible to increase the numbers captured to that extent.

In the revised Simbambi 4.1 version of the population dynamics simulation program, the capture numbers are given by the user and the operational error is not considered, but the demographic probability fluctuation and environmental probability fluctuation are considered.

## 9.4 Investigation for Only One Year Is Not Enough

Survival and reproductive rates among wildlife are not the same every year, and during “strong year classes,” once in several years, wild animals leave many offspring, but the numbers may also be greatly decreased due to “catastrophe” every few years. In the case of Japanese deer, the numbers increased too much resulting in mass death in such spatially limited habitats as the island of Nakajima in Lake Toya.

During a year of extraordinarily heavy snowfall in the Meiji era, moreover, deer did not survive the winter, and mass death was recorded across the whole of Hokkaido. In the last 30 years of the twentieth century, however, no mass death occurred. One reason was the reduced frequency of heavy snowfall, but another reason may have been that expanded afforestation increased mixed forests of conifers and broad-leaved trees, allowing deer to survive the winter even in a year of heavy snowfall.

To return to the topic of fish population management, another example worth noting with regard to annual fluctuation is that of char. These fish are listed as an endangered species because the population is divided into separate groups due to river repair works including dams. In the case of white-spotted char (*Salvelinus leucomaenis*), part of the population goes down to the sea and returns. However, if dam-like structures are constructed downstream, the char are prevented from going upstream (Morita and Yokota 2002). It is a problem if the population is maintained by the breeding of individuals remaining in the rivers. The recruitment rate fluctuates every year, and the survival rate to 1 year old fluctuates from 2 to 15 %. If the estimation of the survival rate is correct, the population size increases when the survival rate is 5 % or higher, and it tends to decrease when the rate is lower than 5 %.

If it is assumed that the survival rate of the year investigated continues every year, future predictions may differ greatly from reality. If the recruitment rate (product of reproduction rate and initial survival rate to the age at recruitment, usually 1 year old) is underestimated, it will be assessed that the population of this species of char will become extinct. This will be the case whether or not new fisheries business is conducted in the location. If the recruitment rate is overestimated, it will be assessed that the survival of the population will not be affected even if some load is imposed by the business.

But both assessments are improper. As indicated earlier, reproductive rates and initial survival rates for wildlife often fluctuate every year. It is impossible to estimate reproductive rates and survival rates by investigating only one year, as is usually the case in environmental impact assessments. Even if a researcher conducts an investigation for five years, quantitative assessment may not be possible.

What should we do then? When assessing environmental impacts, as well as wildlife population levels, it is important to conduct not only a one-year investigation, but also to undertake subsequent post-investigation and reviewing, and other follow-up. Where population increases or decreases are concerned, various conservation measures should be prepared in advance, and the appropriate measures should then be taken in response to events as they evolve. This is the idea of adaptive management described above.

## 9.5 Explanation of Formulae

1. Relationship between stock biomass and its surplus production according to MSY theory—Fig. 9.1

   The ordinary differential equation related to time t was made as follows:

$$dN/dt = r(1 - N/K)N - C \tag{9.1}$$

where $N$ is stock, $C$ catches, and $r$ and $K$ are positive constants called intrinsic rate of natural increase and carrying capacity, respectively. We assumed $r=0.2$ and $K=1,000$.

2. Example of numeric calculation of adaptive marine resource management—Fig. 9.3

$$N(t+1)=\left[N(t)-C(t)\right]\exp\left[r(t)-k\{N(t)-C(t)\}\right] \tag{9.2}$$

where $k$ is a positive constant indicating the magnitude of density effect on the recruitment rate. $r(t)$ is not constant, but is given by:

$$r(t)=\rho r(t-1)+(1-\rho)\ r^*\left[1+\sigma_r\varepsilon_r(t)\right], \tag{9.3}$$

where ρ is the magnitude of autocorrelation in $r(t)$, $r^*$ is a positive constant implying the average$r$; $\sigma_r$ and $\varepsilon_r$ are the magnitude of error in the recruitment process and the random variable between −1 and +1.

The exploitation rate $F(t)$ is determined by:

$$F(t)=\begin{cases} 0\ if\ \tilde{N}(t)<N_{ban} \\ F_{target}\dfrac{\tilde{N}(t)-N_{ban}}{N_{limit}-N_{ban}}\ if\ N_{ban}<\tilde{N}(t)<N_{limit}\ F \\ F_{target}\ if\ \tilde{N}(t)>N_{limit} \end{cases} \tag{9.4}$$

where $\tilde{N}(t)$ is the estimated stock biomass. We give it as:

$$\tilde{N}(t)=N(t)\left[1+\sigma_N\varepsilon_N(t)\right] \tag{9.5}$$

where $\sigma_N$ and $\varepsilon_N$ are the magnitude of estimation error and the random variable between −1 and +1. Catch in year $t$, denoted by $C(t)$, is given by:

$$C(t)=N(t)F(t)\left[1+\sigma_F\varepsilon_F(t)\right] \tag{9.6}$$

where $\sigma_F$ and $\varepsilon_F$ are the magnitude of operational error in fishery and the random variable between −1 and +1.

We set $r=0.2$, $K=N(0)=1,000$.$\varepsilon_N$(t), $\varepsilon_F$(t) and $\varepsilon_r$(t) are uniform random numbers between −1 and +1, and $\sigma_N$, $\sigma_F$, and $\sigma_r$ were assumed to be 80, 30, and 10 %, respectively, in Fig. 9.3c, d. ρ was assumed to be 0.7 in Fig. 9.3c, d. In the case of Fig. 9.3a, they were all made 0.

3. Concept of Japanese deer protection management—Fig. 9.4

The population size of the male and female deer of $i$ years old in year $t$ are denoted by $N_i^{♀}$(t) and $N_i^{♂}$(t), respectively, and $i=0$, 1, 2, 3, and 4+; those aged 4 or older are handled together for simplicity (this is different from the Simbambi program). The population size of the following year is expressed by the following formulae.

$$N_0^{♀}(t)=N_0^{♂}(t)=\ \Sigma\left[N_i^{♀}(t)m_i\right]/2 \tag{9.7}$$

**Table 9.1** Parameter values used for Fig. 9.4

| Age $i$ | 0 | 1 | 2 | 3 | 4 |
|---|---|---|---|---|---|
| Reproduction rate $m_i$ (%) | 0 | 0 | 20 | 40 | 45 |
| Selective catch coefficient $f_i$ | 0.6 | 1 | 1 | 1 | 1 |
| Survival rate of males $p_i^{\female}$ (%) | 70 | 86 | 90 | 93 | 93 |
| Survival rate of females $p_i^{\male}$ (%) | 70 | 82 | 88 | 90 | 90 |

$$N_0^{\male}(t+1) = p_i^{\male}[N_i^{\male}(t) - C_i^{\male}(t)] \tag{9.8}$$

$$N_i^{\female}(t+1) = p_i^{\female}[N_i^{\female}(t) - C_i^{\female}(t)] \tag{9.9}$$

$$N_i^{\male}(t+1) = p_i^{\male}[N_i^{\male}(t) - C_i^{\male}(t)] \tag{9.10}$$

$$N_{4+}^{\female}(t+1) = p_4^{\female}[N_3^{\female}(t) + N_{4+}^{\female}(t) - C_3^{\female}(t) - C_3^{\female}(t)] \tag{9.11}$$

$$N_{4+}^{\male}(t+1) = p_4^{\male}[N_3^{\male}(t) + N_{4+}^{\male}(t) - C_3^{\male}(t) - C_{4+}^{\male}(t)] \tag{9.12}$$

$$C_i^{\female}(t) = C^{\female}(t)N_i^{\female}(t)f_i \,/\, \Sigma\,[N_j^{\female}(t)f_i] \tag{9.13}$$

$$C_i^{\male}(t) = C^{\male}(t)N_i^{\male}(t)f_i \,/\, \Sigma\,[N_j^{\male}(t)f_i] \tag{9.14}$$

Where $m_i$ is pregnancy rate of $i$ year-old, $f_i$ is selective catch coefficient of $i$ year-old, $C^{\female}(t)$ and $C^{\male}$(t) are total numbers of males and females captured in year $t$, respectively. We assumed selective catching considering a high possibility of not capturing sub-adult animals. It was assumed that natural death would occur for individuals after they were captured and reduced in number (Table 9.1).

## 9.6 Conclusion

We need the concept of risk in fisheries and wildlife management because of several kinds of uncertainties. Traditional theory of maximum sustainable yield ignores such uncertainties and is not convincing for fishery operators, as biomass of fishery resources usually fluctuates remarkably. Risk analysis using stochastic models is useful and more convincing, even though the true mechanism of population dynamics is still unknown.We need risk evaluation to show a prediction range with some allowance.

I recommend adaptive management as a method of coping with such uncertainty. Adaptive management can change the exploitation rate of fish/wildlife populations. It can reduce future uncertainties compared to a policy with a constant exploitation rate. Therefore, adaptive management is indispensable in fisheries and wildlife management. We do not make a single prediction, but show the risk of management failure, which is rarely eliminated because of uncertainties. We compare the risk of management failure between two scenarios in order to show the advantages of a recommended management policy.

## References

Clark CW (1990) Mathematical bioeconomics: the optimal management of renewable resources, 2nd edn. Wiley, New York

Holling CS (1966) The functional response of invertebrate predators to prey density. Mem Entomol Soc Can 48:1–86

Kawai H, Yatsu A, Watanabe C, Mitani T, Katsukawa T, Matsuda H (2002) Recovery policy for chub mackerel stock using recruitment-per-spawning. Fish Sci 68:961–969

Matsuda H, Abrams PA (2004) Effects of predator–prey interactions and adaptive change on sustainable yield. Can J Fish Aq Sci 61:175–184

Matsuda H, Abrams PA (2006) Maximal yields from multi-species fisheries systems: rules for systems with multiple trophic levels. Ecol Appl 16:225–237

Matsuda H, Abrams PA (2013) Is feedback control effective for ecosystem-based fisheries management? J Theor Biol 339:122–128

Mori M, Katsukawa T, Matsuda H (2001) Recovery plan for the exploited species: southern bluefin tuna. Pop Ecol 43:125–132

Morita K, Yokota A (2002) Population viability of stream-resident salmonids after habitat fragmentation: a case study with white-spotted char (*Salvelinus leucomaenis*) by an individual based model. Ecol Model 155:85–94

Walters CJ (1986) Adaptive management of renewable resources. McMillan, New York

# Chapter 10
# Valuation of Non-Marketed Agricultural Ecosystem Services, and Food Security in Southeast Asia

**Ryohei Kada**

**Abstract** Food security, closely linked with environmental issues, has become one of the most important issues in the twenty-first century. In recent decades especially, ecological degradation has been spreading, and is negatively affecting food supply and food safety conditions in many Southeast Asian countries. Such degradation can include sedimentation, reduced water quality, and frequent flood occurrence, many of which are enhanced by climate change impacts. Based on an international research project conducted by RIHN (Research Institute for Humanity and Nature) in collaboration with Yokohama National University and the University of the Philippines, we discuss the recent changes in food risks and the factors contributing to expansion of such risks in Southeast Asia. Our study demonstrates that non-marketed ecosystem services from sustainable agricultural land use can provide significant economic value, and developing a mechanism to pay for ecosystem services is crucial in enhancing sustainable agricultural development.

**Keywords** Ecological degradation • Ecosystem services • Resource management • Southeast Asia • Valuation of non-marketed services

## 10.1 Food Security Issues in Southeast Asia

Many Asian countries increased agricultural and food production in the course of their rapid economic development, leading to extensive degradation of natural resources and the reduction of ecosystem services. Such unsustainable patterns of development gradually endanger biodiversity and ecosystem services that are crucially important in Southeast Asia.

---

R. Kada (✉)
Research Institute for Humanity and Nature (RIHN), Kyoto, Japan

Graduate School of Environment and Information Sciences,
Yokohama National University, Yokohama, Japan
e-mail: kada@ynu.ac.jp; kada@chikyu.ac.jp

N. Kaneko et al. (eds.), *Sustainable Living with Environmental Risks*,
DOI 10.1007/978-4-431-54804-1_10,  

The basic issue here is whether and how Asian agriculture and fisheries can supply a sufficient amount and quality of food stably to meet the needs of the rapidly growing population without jeopardizing natural and environmental resources.

Probably the most critical impacts of resource degradation are on food security for the poorer populations in rural Asia. On the other hand, environmental regulations to combat such resource degradation might increase the cost of production and decrease farmers' net returns for some commodities in the short-run. Lower returns might also alter the pattern of agricultural products and initiate new directions in domestic production and international trade.

It has been estimated that food production must rise 50 % to feed nine billion people by 2050 (FAO 2008). The increased population growth accompanied by significant land use change will substantially reduce the availability of non-marketed ecosystem services, threaten agricultural sustainability, rapidly alter the global climate, and soon put human well-being at serious risk.

Can we take up the global challenge? This study demonstrates that taking some steps at a local level may help significantly in ensuring agricultural sustainability, helping to mitigate climate change, and maintaining human well-being at a global level. Such steps should include recognizing and estimating the value of non-marketed ecosystem services and creating payment for such services.

Agriculture supplies ecosystem services including marketed goods such as food and fiber, and non-marketed services such as flood regulation, disease control, water purification, and scenic views. People generally assume that the non-marketed ecosystem services, which are not traded in traditional markets, have no economic significance (Daily and Ellison 2002). In recent years, agriculture in Asia especially has experienced significant land use change and soil degradation due to urbanization, non-agricultural practices, deforestation, and overexploitation of agricultural resources as a result of population growth. Consequently, the level of non-marketed ecosystem services has drastically reduced, giving rise to ecological risks. Studies estimating the values of non-marketed ecosystem services that emerge from agriculture are still inadequate, and consequently, the decision-making process involving agricultural sustainability and food security has remained flawed.

Given this, our study demonstrates that non-marketed ecosystem services from agriculture can have significant economic value, and developing a mechanism to pay for ecosystem services is crucial in enhancing sustainable agricultural development. Furthermore, failing to accommodate non-market values in the decision-making process for agricultural development can even threaten agricultural sustainability and human well-being by generating considerable externalities and ecological risks, as well as market failures.

## 10.2 Need for Valuing Non-Marketed Ecosystem Services

In this study, we have used the classification of ecosystem services devised by Millennium Ecosystem Assessment (Millennium Ecosystem Assessment (MEA) 2005) (Fig. 10.1), and illustrated a number of valuation methods for non-marketed

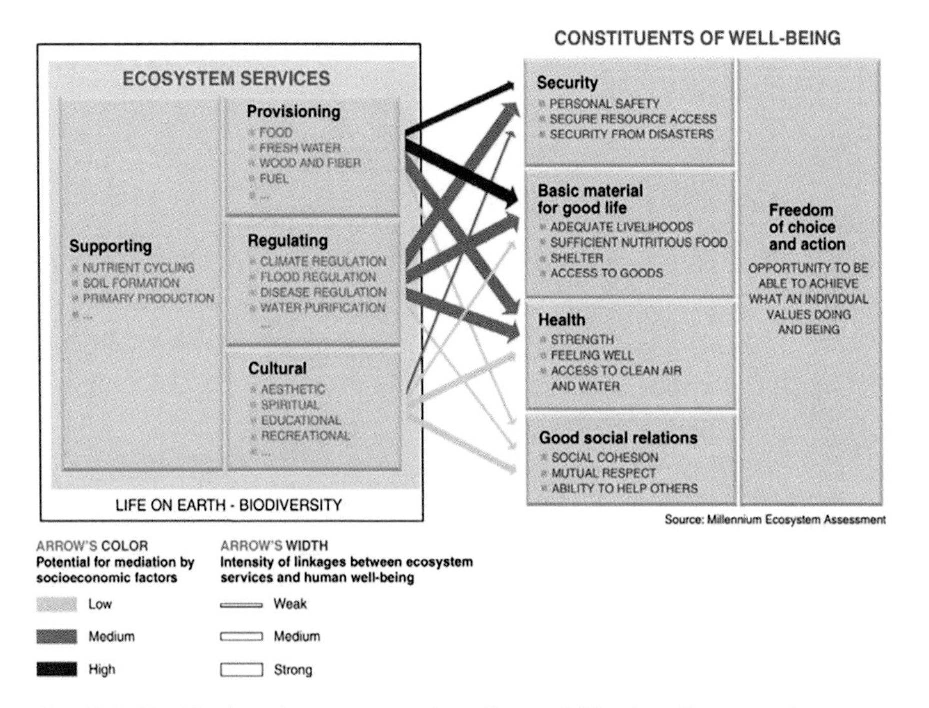


**Fig. 10.1** Classification of ecosystem services (*Source*: Millennium Ecosystem Assessment (MEA) (MEA) (2005))

ecosystem services stated in Pearce and Howarth (2000). This classification includes supporting, provisioning, regulating, and cultural services. The supporting services maintain the other three, which are linked to human well-being. In line with this, we have illustrated international examples of payment for ecosystem services and demonstrated how estimating the economic value of ecosystem services and developing a mechanism for payment involving beneficiaries and providers could help address issues associated with economic externalities, ecological risks, and market failures, and subsequently, enhance human well-being. Ecosystem services, which are defined as the benefits provided from nature to humans, include supporting services such as primary production and soil formation, provisioning services such as food and water, regulating services such as flood and disease control, and cultural services such as recreational benefits (Millennium Ecosystem Assessment (MEA) 2005) (Fig. 10.1). Supporting services maintain the other three services.

Human actions, externalities, ecological risks, and human well-being are well interlinked. This means that when human actions reduce the level of ecosystem services, ecological risks emerge as a distribution of potential negative externalities. An externality is generally defined as an action of an individual or a group of individuals that inflicts a cost or benefit on others. A human action can have both negative and positive impacts on others. For example, when a farmer uses an excess amount of chemical fertilizers on his or her land, the nearby water body becomes

**Table 10.1** Degradation of ecosystem services and ecological risks in Asia

| Reduction in ecosystem services (ES) | Ecological risks due to a reduced level of ES | Human activities reducing ES | Countries in Asia where ecological risks are highly likely |
|---|---|---|---|
| *Supporting*<br>• Nutrient cycling<br>• Soil formation<br>• Primary production | Poor soil quality<br>Destruction of genetic resources such as fish, frogs, and earth worms in paddy fields | Use of chemical fertilizers<br>Use of chemical insecticides<br>Introduction of aquaculture | Bangladesh, India, Indonesia, Nepal |
| *Provisioning*<br>• Food<br>• Fresh water<br>• Wood and fiber<br>• Fuel | Food insecurity<br>Water pollution<br>Soil erosion | Conversion of agricultural land for human settlements<br>Deforestation | Bangladesh, India, Nepal, Indonesia, the Philippines, Sri Lanka |
| *Regulating*<br>• Climate regulation<br>• Flood regulation<br>• Disease regulation<br>• Water purification | GHG emissions<br>Flooding<br>Public health hazards<br>Water Pollution | Extension of oil palm and sugar cane plantations | Indonesia, Malaysia, the Philippines |
| *Cultural*<br>• Aesthetic<br>• Spiritual<br>• Educational<br>• Recreational | Reduced aesthetic appeal<br>Loss of recreational amenities<br>Less communication among people | Use of chemical fertilizers<br>Use of chemical insecticides | Bangladesh, India |

polluted and biomass resources will be gradually lost due to the reduction in ecosystem services such as water purification. However, there is no obligation to compensate for loss of ecosystem services; consequently, the farmer does not pay for the damage that he or she inflicts on the owner of the water body. Similarly, there is no requirement for the owner of the water body to pay the farmer (as the owner of the land) if he or she abstains from using chemical fertilizers (the absence of which could inhibit the farmer's production).

As shown in Table 10.1, human activities such as use of chemical fertilizers and insecticides, and conversion of agricultural land for human settlements, can generate ecological risks. These include poor soil quality, water pollution, and food insecurity resulting from reduced ecosystem services, especially provisioning services.

In Indonesia serious land use changes occur as a result of rapid expansion of oil palm plantations that replace ecologically valuable tropical rainforest and mangrove. This is a situation in which human action causes $CO_2$ emission responsible for climate change and global warming. Furthermore, clearance and draining of peatland enhances fire and causes carbon stock to emit $CO_2$ into the atmosphere. Although the country's economic development is essential, therefore, the ecological risks associated with the development are pervasive.

Likewise in Malaysia land use changes occur due to expansion of palm oil production, and the changes exacerbate crises already caused by deforestation, ecosystem degradation, and biodiversity loss. The country currently faces serious issues involving the trade-off between oil-palm growth and maintenance of ecosystem services. Furthermore, many ecological risks have emerged in the agricultural sector, such as water supply shortages, soil erosion and fertility loss, water-logging effects, salinity, and mangrove conversion. In order to manage these ecological risks, the country instituted a range of legal and regulatory measures such as the National Environmental Policy, the National Conservation Strategy, the National Policy on Biological Diversity, and the Environmental Quality Act.

In the Philippines, human-induced ecological risks occur due to land use changes in wetlands. Wetlands are important for both humans and birds. Although particular areas are set aside as bird sanctuaries by local governments with cooperation from local communities, there is growing pressure for cultivated wetland areas to meet the increasing needs of humans for food and fiber. To increase food production in cultivated areas more water is required, and this aggravates water pollution and soil erosion, while reducing carbon sequestration. Consequently, the pressure on cultivated areas reduces the level of ecosystem services. Like other countries, the Philippines is trying to balance agricultural production with supply of ecosystem services, but efforts to do so are affected by typhoons, drought, floods, and social unrest. To address the issues of ecological risks, the country has introduced a wide range of environmental laws and policies on land use management and agrarian reforms, with greater priority placed on community-based management. However, policies related to national land use are not on the agenda and this has caused confusion and conflicts on the part of implementing agencies and authorities.

Recent expansion of bio-energy production, which has occurred indiscriminately and on a major scale, has resulted in new environmental obstacles, especially in Indonesia, Malaysia, and the Philippines. Typically, the expansion of oil palm, coffee, or banana plantations has caused massive deforestation and a loss of biodiversity that could include precious genetic resources found only in tropical Asia. Ironically, $CO_2$ emission in plantation crops occurs at a considerably higher rate than in the conventional crops and grains.

Goda et al. (2006) have listed a number of negative externalities that ultimately result from human activities such as degradation, livestock production, and use of pesticides in agriculture. Due to population growth and rapid economic development, many Asian countries have increased agricultural food production. However, these human activities have resulted in extensive degradation of natural resources and the reduction of ecosystem services. In particular, the adoption of modern technology including intensive use of fertilizers and monoculture has caused serious soil and water degradation. For example, substantial soil degradation has resulted from deforestation (30 %), overcultivation (32 %), and overgrazing (26 %) occurring due to population growth and economic development in the Asian region.

In Japan, the decrease in farmland area has caused various social problems. Due to both rapid urbanization and abandonment of farmland, ecosystem services from farmlands have been lost, resulting in more frequent occurrences of flooding in the surrounding areas and city areas downstream. Although chemical fertilizer input is an essential element of enhancing productivity, an excessive use of chemical inputs in the long run has resulted in negative externalities such as water pollution and biodiversity loss.

Groundwater contamination due to crop and livestock production activities has become an important public issue in some areas of Japan. Nitrates are a major source of water contamination, introduced not only through the use of chemical fertilizers, but also associated with the amalgamation of animal manure.

Similarly, pesticide residues can be toxic to humans and animals in food, feed, or drinking water. They also may be hazardous to farm workers during application and have undesirable side effects on non-targeted living organisms within the natural ecosystem.

It is of considerable importance to develop a mechanism to pay for ecosystem services involving the beneficiaries and producers of such services in the agricultural sector. This could address the externality and market failure issues in agriculture, and consequently, increase food production in a sustainable way.

Intensive livestock production systems, especially confinement livestock operations, accumulate large amounts of animal waste in solid and liquid forms. Manure, in concentration or untreated, can be considered a potential hazard to the environment due to its high content of nitrate, ammonia, phosphate, and potassium. As manure decomposes, various microelements (zinc, copper, manganese, iron, etc.) are released directly into surface rivers and lakes, or are leached through the soil to groundwater sources. Other things being equal, high animal densities indicate strong potential for pollution problems that could translate into higher costs of production if waste management costs rise. Such high densities are frequently associated with technological developments found in modernized, large-scale livestock operations, as well as in China and other developing countries.

What all these consequences imply is that we need to estimate the values of non-marketed ecosystem services and develop a mechanism to pay for them in order to address such issues.

## 10.3 Issues of Valuation and Payment for Non-Marketed Ecosystem Services

Estimating the economic values of non-marketed ecosystem services is a challenging issue. There are a number of approaches available to estimate the values, and the selection of a particular approach is problematic. In addition, such estimation requires significant knowledge regarding the provision, distribution, and valuation of non-marketed ecosystem services. Nonetheless, it is of considerable benefit to develop some form of valuation approach.

Figure 10.2 demonstrates that many approaches exist to estimate economic values of environmental goods and services, as illustrated by the *satoyama* and *satoumi* cases. Total Economic Value (TEV) comprises revealed (conventional market) and stated (hypothetical market) preferences.

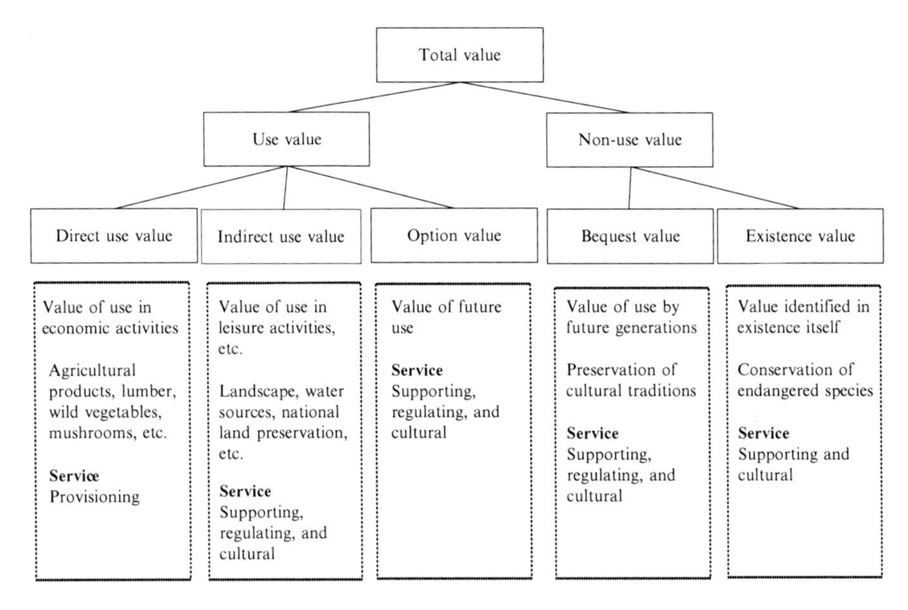


**Fig. 10.2** Typology of economic values and relevant ecosystem services of *satoyama* and *satoumi* (Kada et al. 2012)

### *10.3.1 Revealed Preferences (Conventional and Surrogate Markets)*

Revealed preferences methods use existing markets for relevant goods and services with a view to revealing TEV of environmental qualities. For most regular, tradable goods in a market, the use-values are significant and TEV is determined principally by the use value of the goods as revealed by their market price. However, the non-use value may be highly significant because people may value an environmental good for its existence and capacity to support sustainable livelihoods for current and future generations. The major drawback of the revealed preference method when compared to the stated preference method is its inability to accommodate and measure these non-use values.

Two of the commonly practiced methods in the revealed preferences group are Hedonic Pricing and Travel Cost Methodologies.

1. *Hedonic Pricing Method (HPM)*: A hedonic pricing approach is used to estimate economic values of ecosystem services (i.e., benefits from nature) that affect market prices. The approach involves the quality of the environment and ecosystem (such as the degree of air pollution and noise) and environmental amenities (e.g., aesthetic views, or distance to the central business district, recreational sites such as botanical gardens, and hospitals). The Hedonic Pricing Method (HPM) assumes that differences in amenities for the housing environment are

reflected in land prices and wages. By this hedonic pricing approach it is estimated that Japan's paddy field amenities are worth approximately JPY 12 trillion.

2. *Travel Cost Method (TCM)*: The travel cost method or approach is a way to estimate economic values of recreational sites. This approach is used to analyze nature-based recreation demand and estimate economic values of the natural resources. The TCM is widely used to estimate the value of recreational sites and could be used to estimate willingness to pay (WTP) for ecotourism to visit world heritage sites such as the Ifugao Rice Terraces in the Philippines.

   In Concepcion (2006) we find that a socio-economic team used the TCM to evaluate the agri-tourism function of agriculture in mango orchards in the Philippines. They used the formula: Valuation = (number of domestic and foreign tourists) × (average opportunity cost of work + average cost per visit + average value of mangoes and souvenir items bought), where the number of tourists is 13,322, average travel cost is USD 36.62, average opportunity cost is USD 18.69, average accommodation cost is USD 14.79, and average cost of mangoes and other souvenir items is USD 23.42.

### *10.3.2 Stated Preferences (Hypothetical Preferences)*

The stated preferences approach is primarily survey-based and depends largely on what the respondents say rather than what they do. The approach posits a hypothetical market or payment scenario involving unpriced environmental goods and services, asks respondents to place a dollar value on goods and services, and draws conclusions about respondents' willingness to make payment for some specific attributes of the goods and services.

The stated preferences group includes contingent valuation, and conjoint analysis.

1. *Contingent Valuation (CV)*: CV was developed. The main purpose of CV is to estimate people's Willingness to Pay (WTP) or Willingness to Accept (WTA) based on *a single feature* (rather than multiple features) of an environmental good. This approach asks people to state their willingness to pay, and is contingent on a specific hypothetical market scenario.
2. *Conjoint Analysis (CA)*: CA is a statistical technique used to find out how people value *multiple features* of an individual environmental good.

CA may include Contingent Ranking, Contingent Rating, Paired Comparison, and Choice Modeling which are closely related approaches to environmental valuation.

The most important ecosystem service from agriculture is perhaps food security, while we must also maintain agricultural sustainability. Due to soil degradation and water pollution, small farmers and poor city dwellers encounter poverty and food insecurity all over the world.

Kada (2012) identified the potential and challenges in developing payment for ecosystem services in Japan, as follows:

- An increasing number of individuals and groups from cities are willingly assisting farmers and rural communities with their work, while at the same time enjoying the rural amenities.
- The Japanese Ministry of Agriculture, Forestry and Fisheries (MAFF) has introduced a direct payment scheme for farmers and rural communities to maintain the existence of paddy fields.
- To create ecologically-friendly wetlands, parks, and ponds in harmony with agricultural activities, attempts have been made on both a public and a voluntary basis to undertake groundwork activities and create trusts to conserve the landscape and preserve genetic resources (e.g., insects, fish, birds, and flora).
- Consumers and producers are affiliating to support regional agriculture through contracts between the two parties for fresh and safe farm products. Direct marketing systems such as farmers' markets and producer-consumer affiliated marketing have become very popular in Japan today.
- There has been an expansion in green tourism, whereby city dwellers are invited to spend their leisure time visiting rural landscapes, enjoying farming experiences, farming for their own consumption, and voluntarily working in forests and paddy fields in remote rural areas.
- Local governments are providing direct support for environmentally-friendly farming through direct subsidies to farmers who keep paddy fields for flood mitigation purposes. To be eligible for this subsidy, the local government requires that the owners of paddy fields sign a contract to maintain the paddy land or farmland.

As shown in Fig. 10.2, non-marketed ecosystem services derived from agricultural and rural areas can be measured. The table measures regulating services, such as prevention of floods, soil erosion, and landslides, as well as cultural services, such as recreation and relaxation.

What this implies is that estimating economic values for non-marketed ecosystem services is possible, if imprecise. What is more, estimating economic values is crucial in creating a market for ecosystem services (payment for ecosystem services). Both non-coercive government intervention and cooperation from the natural resource users would appear to be essential in establishing such a market. In order to maintain the level of non-marketed ecosystem services provided by agriculture, therefore, it is important to assist farmers with new technology that requires a reduced amount of chemical fertilizers and insecticides, reduces the conversion of agricultural land use into non-agricultural, and compensates farmers for conserving agricultural land.

## 10.4 Conclusion

Although food production must increase to feed the burgeoning population, it is possible to maintain agricultural sustainability to an extent by estimating the values of non-marketed ecosystem services and creating a market for such services (i.e., a mechanism to pay for ecosystem services). This will both maintain the availability of ecosystem services and enhance human well-being.

Agriculture provides not only marketed ecosystem goods (e.g., food and fiber), but also non-marketed ecosystem services (e.g., water purification, disease control, and scenic views). People have usually ignored the economic significance of the ecosystem services and considered them to be unlimited. However, as the overexploitation of agricultural resources has reduced the level of non-marketed ecosystem services and diminished human well-being, people have started to recognize that such ecosystem services are finite, and have economic values. They have also understood that valuing these ecosystem services is crucial in creating a market for them.

Past studies recognized that non-marketed ecosystem services have economic values, and developed a number of approaches to estimate those values, but making precise estimates still remains a challenging task. Although the estimates are far from precise, however, the recognition of value in itself has profound policy implications in terms of enhancing food security and making agricultural development more sustainable by addressing externalities, ecological risks, and market failures. Thus, taking up appropriate action at the local level will help us counter the global challenges.



## References

Concepcion RN (2006) Emerging roles of land and water management and agri-tourism to enhance the multifunctionality of agriculture in the Philippines. Country case studies on multifunctionality of agriculture in ASEAN countries. The ASEAN Secretariat, Ministry of Agriculture, Forestry and Fisheries of Japan, Jakarta

Daily GC, Ellison K (2002) The new economy of nature: the quest to make conservation profitable. Shearwater Books/Island Press, Washington, DC

FAO (2008) Ecosystem services sustain agricultural productivity and resilience. Plant Production and Protection Division. Food and Agriculture Organization, Rome

Goda M, Kada R, Yabe M (2006) Concept and significance of multifunctionality of agriculture: a global perspective. Country case studies on multifunctionality of agriculture in ASEAN countries. The ASEAN Secretariat, Ministry of Agriculture, Forestry and Fisheries of Japan, Jakarta

Kada R (2012) Opportunities and challenges for rebuilding and effective use of *satoyama* resources. Glob Environ Res 16(2):173–179

Kada R, Kohsaka R, Saito O (2012) Why is change to satoyama and satoumi a concern? In: Duraiappah AK et al (eds) Satoyama-satoumi ecosystems and human well-being. United Nations University Press, Tokyo/New York

Millennium Ecosystem Assessment (MEA) (2005) Ecosystems and human well-being: synthesis. Island Press, Washington, DC

Pearce DW, Howarth A (2000) Technical report on methodology: cost benefit analysis and policy responses. RIVM report

# Chapter 11
# Emerging Socio-Economic and Environmental Issues Affecting Food Security: A Case Study of Silang-Santa Rosa Subwatershed

**Roberto F. Rañola Jr., Fe M. Rañola, Maria Francesca O. Tan, and Ma. Cynthia S. Casin**

**Abstract** The Silang-Santa Rosa subwatershed feeds into Laguna Lake to the south-east of Metro Manila. This case study of the subwatershed provides some insights on the interactions between people and institutions within their given natural environment. The link between the socio-economic conditions of households and the quality of the ecosystem resources available is premised on the extent to which socio-economic conditions influence household decisions relating to the use and management of land and water resources. The major issue is how people might be able to improve, protect, and expand their current resource base or level of acquirement given the different types of risks they face. In the subwatershed's upstream areas, the risk is from soil degradation coupled with inefficient farm production systems that lead to low farm productivity. In the downstream areas, households face poverty-related issues such as food insecurity and low income from declining fishery resources. In addition, pollution from both upstream and downstream areas threatens their livelihoods and increases the incidence of water-borne diseases. To address these issues, it is proposed that the different stakeholders could be enlisted to develop an integrated development plan that would reflect their common interests and vision for the watershed area feeding into Laguna Lake. The plan should deal with major issues such as land and water degradation, poverty, livelihoods, food security, health, farm production efficiency, and marketing systems, as well as regulatory and economic instruments that would reduce land and water degradation.

**Keywords** Farm income • Food security • Global environmental change

R.F. Rañola Jr. (✉) • F.M. Rañola • M.F.O. Tan
Department of Agricultural Economics, College of Economics and Management,
University of the Philippines Los Baños, Los Baños, Philippines
e-mail: bert1866@gmail.com; fkm1866@yahoo.com; france.tan.315@gmail.com

M.C.S. Casin
Forestry Development Center, College of Forestry and Natural Resources,
University of the Philippines Los Baños, Los Baños, Philippines
e-mail: chingcasin@yahoo.com

N. Kaneko et al. (eds.), *Sustainable Living with Environmental Risks*,
DOI 10.1007/978-4-431-54804-1_11, © The Author(s) 2014

## 11.1 Introduction

Food systems and food security are greatly affected by environmental changes such as land degradation, loss of biodiversity, and alteration in hydrology or climate patterns. These changes may be either natural or anthropogenic in origin, although it is the alterations of anthropogenic origin that contribute most to environmental change. Food systems include activities such as food production, processing, packaging, and distribution and retailing. They are complex systems influenced by environmental, social, and economic factors, and they link directly with policy and public health issues.

Currently, the accepted definition of food security as defined by the International Conference on Nutrition (ICN) is "access by all people at all times to the food needed for a healthy life," [FAO/WHO 1992a, as cited by the Committee on World Food Security of the Food and Agriculture Organization (Food and Agriculture Organization-Food and Nutrition Division and Publishing Management 1997)]. Food insecurity, on the other hand, is defined as lack of access to enough food for a healthy and active life. According to a definition that had its origins in the 1970s, however, food security refers to the *availability*, *accessibility*, *utilization*, and *stability* of food supplies (Stamoulis and Zezza 2003, as cited in Springer Editorial 2009).

Food *availability* is often associated with climatic changes that can affect the agriculture, aquaculture, and forestry sectors. *Accessibility*, however, depends on the knowledge and resources of households themselves. Such knowledge and resources include strategies utilized by households for food acquisition, that is, their ability to produce food and exchange assets for food. The strategies may include borrowing money, selling assets, or engaging in wage labor to provide food for the family. While neither rural nor urban households are spared from food insecurity, especially during seasonal troughs, the former are able to produce their own food, while the latter purchase most of their foods. The urban households are therefore more affected by food insecurity, especially if jobs are lost, incomes fall, and food prices increase. In terms of *utilization*, women are usually responsible for food procurement, preparation, and storage, especially in rural areas. *Stability* of food supply, meanwhile, is defined by use of a set of strategies to cope with both major and minor stresses. Communities that live in stressful environments, in particular, have developed strategies to reduce the impact of environmental stresses on household food security in the short- and/or long term. An important strategy is protection of the resource base through production practices that do not damage the environment to the detriment of the agriculture, fishery, and forestry sectors. An example would be changing cropping systems to increase farm productivity. Another coping strategy is to ensure that the procurement of food does not lead to the loss of a household's capacity to produce using its assets such as land, equipment, and farm animals.

The issue of risks to food security from environmental degradation is of major importance, because the poor are the ones most affected. In rural areas where a high incidence of poverty exists, the rural poor are forced to cultivate fragile and marginal lands causing soil erosion. This renders them very vulnerable to environmental risks such as the diminution and contamination of their water supplies, and reduction in farm productivity that has a direct impact on their food security (Food and Agriculture Organization-Food and Nutrition Division and Publishing Management 1997).

This paper seeks to present the impacts of some socio-economic factors and related environmental risks on food security in the Silang-Santa Rosa subwatershed. A good understanding of this link is a crucial step in developing strategies for addressing food security.

## 11.2 Conceptual Framework

Food security is affected by a number of socio-demographic and environmental conditions in the Silang-Santa Rosa subwatershed (Fig. 11.1). The current environmental condition of the subwatershed is primarily the result of anthropogenic activities extending from the upstream communities to those downstream. Environmental risks that are major threats to food security in the Silang-Santa Rosa subwatershed include land conversion, soil degradation, and consequent soil erosion. An additional threat is the increase in pollution resulting from urbanization and an expanding population. These environmental factors can affect the livelihood of dwellers residing within the community, who are primarily dependent on farming and fishery to make a living. Moreover, since individual sources of livelihood determine the amount of food that can be produced, as well as the additional income that can be generated to purchase food, socio-demographic characteristics such as age and educational attainment also have a major influence on the food security of individuals.

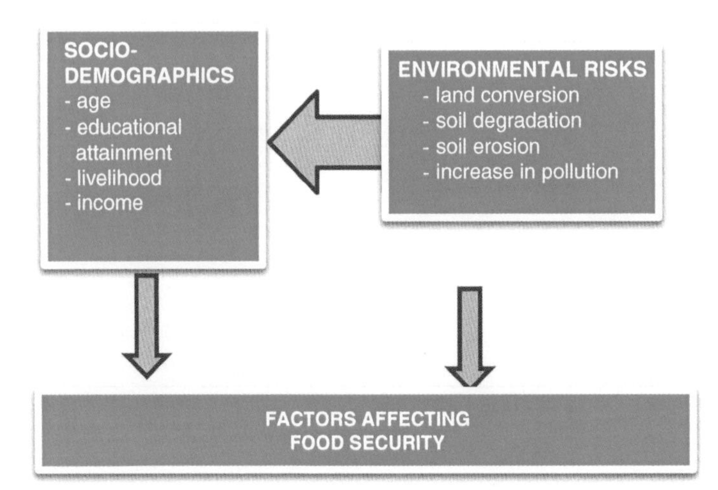


**Fig. 11.1** Conceptual framework of food security in Silang-Santa Rosa subwatershed

## 11.3 Study Site

The research sites were classified into three types of watershed area: upstream (Pulong Bunga), midstream (Tartaria), and downstream (Santo (Sto.) Tomas and Aplaya) (Fig. 11.2). The slopes in the upstream area range from 8 to 18 %.

The *barangay* (district) of Pulong Bunga is an agricultural community located in the upstream municipality of Silang in Cavite Province, near the municipality of Tagaytay. Tartaria, in the midstream area, is another agricultural community in Silang, and portions of its tributaries feed into the downstream municipality of Santa (Sta.) Rosa. According to responses obtained during farmer interviews, there are plans by some private developers to convert the agricultural areas of this *barangay* into built-up areas. The downstream study sites comprise the *barangays* of Sto. Tomas and Aplaya, located in the Laguna Province municipalities of Biñan and Sta. Rosa, respectively. They are regularly flooded and considered to be environmentally critical zones. Fishing is a major source of income in these districts.

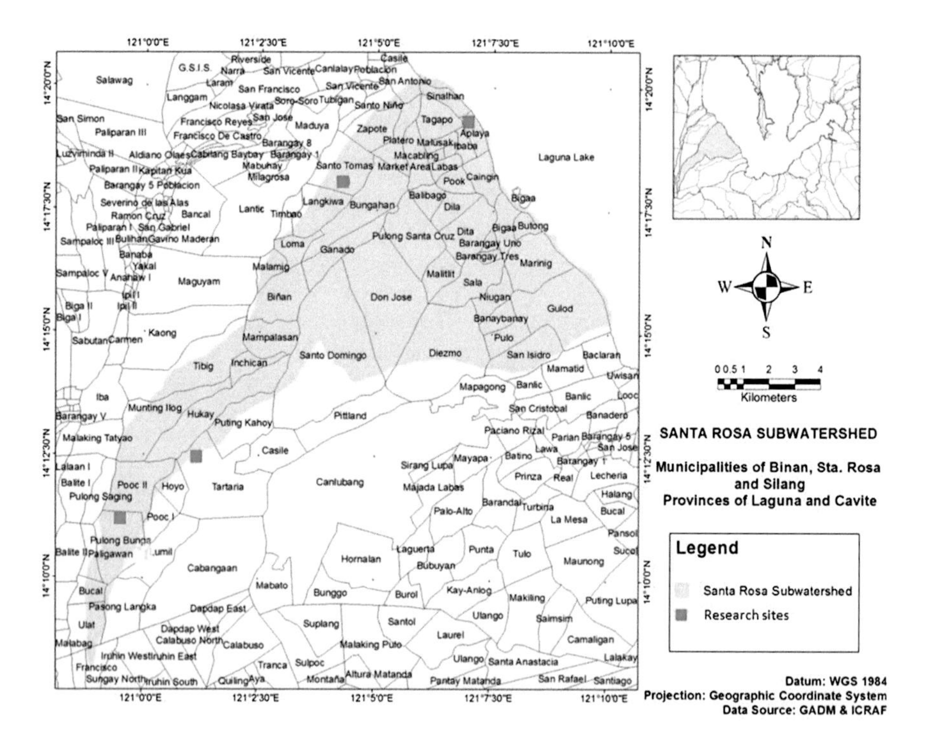


**Fig. 11.2** Map of the Silang-Santa Rosa subwatershed

## 11.4 Results and Discussion

This section provides a summary of household circumstances in the upstream, midstream, and downstream topographical locations within the Silang-Santa Rosa subwatershed. This is important for gaining a good understanding of how environmental risks or factors affect the incidence of poverty and food insecurity in the subwatershed.

### *11.4.1 Socio-Demographic Characteristics*

The socio-demographic characteristics and conditions of farmers and fishermen have a major influence on their vulnerability to natural and anthropogenic hazards, their use and management of land and water resources, and their adaptation strategies to cope with these hazards. This has an effect on their livelihoods, incomes, food security, and health.

Table 11.1 shows the socio-demographic characteristics of households. The upstream and downstream residents of the watershed are relatively homogenous in

**Table 11.1** Household characteristics of respondents surveyed, Silang-Santa Rosa subwatershed, 2013

| | Upstream | Midstream | Upper downstream | Lower downstream |
|---|---|---|---|---|
| Characteristics | Pulong Bunga (n=39) | Tartaria (n=39) | Sto. Tomas (n=42) | Aplaya (n=19) |
| Cultural origin | 77 % Native<br>23 % Migrant | 31 % Native<br>69 % Migrant | 90 % Native<br>10 % Migrant | 100 % Native |
| *Average age* | | | | |
| Household head | 51 | 49 | 42 | 47 |
| Spouse | 42 | 34 | 38 | 45 |
| *Household size* | | | | |
| Range | 2–10 | 2–10 | 2–10 | 3–7 |
| Mean | 5 | 5 | 5 | 5 |
| Type of family | 67 % Nuclear<br>33 % Extended | 74 % Nuclear<br>26 % Extended | 74 % Nuclear<br>26 % Extended | 68 % Nuclear<br>32 % Extended |
| *Education* | | | | |
| Household head | 23 % Primary and Intermediate<br>38 % Secondary<br>38 % Tertiary<br>1 % Postgraduate | 37 % Primary and Intermediate<br>35 % Secondary<br>29 % Tertiary | 30 % Primary and Intermediate<br>59 % Secondary<br>11 % Tertiary | 22 % Primary and Intermediate<br>56 % Secondary<br>22 % Tertiary |
| Spouse | 30 % Primary and intermediate<br>30 % Secondary<br>40 % Tertiary | 27 % Primary and intermediate<br>46 % Secondary<br>27 % Tertiary | 40 % Primary and intermediate<br>56 % Secondary<br>4 % Tertiary | 11 % Primary and intermediate<br>78 % Secondary<br>11 % Tertiary |

their cultural origins. In the lower downstream *barangay* of Aplaya, all residents are native to the district, while in the upper downstream area of Sto. Tomas, and the upstream area of Pulong Bunga, 90 % and 77 %, respectively, are native to their districts. In the midstream area of Tartaria, however, the majority (69 %) are migrants, having migrated within their own municipality from other *barangays* in Silang, or from neighboring Batangas Province.

Most of the respondents are still in their productive years, with average ages of 42 and 47 in the downstream areas of Sto. Tomas and Aplaya, respectively, and average ages of 51 and 49 in Pulong Bunga and Tartaria. Family sizes in Aplaya households are relatively small, ranging from 3 to 7 members, compared to the bigger families of 2 to 10 in Sto. Tomas, Tartaria, and Pulong Bunga.

The literacy rate is high in the study areas. Among household heads in the downstream areas of Sto. Tomas and Aplaya, 59 % and 56 %, respectively, have reached or finished the secondary level of education. In addition, 56 % of spouses in St. Tomas and 78 % of spouses in Aplaya have reached or finished secondary education. In the midstream area, household heads and their spouses have attained higher levels of education, with 29 % and 27 %, respectively, having reached the tertiary level of education. This also holds true in the upstream *barangay*, where 38 % of household heads and 40 % of spouses have reached or earned college degrees.

### *11.4.2 Labor Resources*

The available labor resources can be the most important family asset for the upstream and midstream agricultural households as well as the downstream fishing households. Across sites, the households are predominantly nuclear families whose heads are engaged in farming and fishing activities. The contribution of family labor in farming and fishing, however, is limited. Across sites, the percentages of women that remain at home as housewives are: 40 % of women in upstream areas, 60 % in midstream areas, 62 % in upper downstream areas, and 44 % in lower downstream areas. The rest are either engaged in trading daily goods, selling ornamental plants, or are employed as workers in factories and offices. Other family members who are not gainfully employed provide additional sources of unpaid family labor in farms. To cope with increasing household expenses, the majority of household heads in the upstream and midstream areas have a secondary occupation, with 35 % and 31 % engaged in transport and construction work, respectively. In the downstream sites, meanwhile, only 13 % in Sto. Tomas and 11 % Aplaya, have secondary employment(Table 11.2).

### *11.4.3 Household Financial Status*

In the upstream area, the average annual income for a family of five is PHP 52,508. Given that in 2009 the NSCB (National Statistical Coordination Board 2009) estimated the household poverty threshold at an annual income equal to or less than

**Table 11.2** Employment status of households in Silang-Santa Rosa subwatershed, 2013

| | Upstream | Midstream | Upper downstream | Lower downstream |
|---|---|---|---|---|
| Employment/occupation | Pulong Bunga (n=39) | Tartaria (n=39) | Sto. Tomas (n=42) | Aplaya (n=19) |
| **Household head** | | | | |
| *Primary* | | | | |
| Farmer (%) | 33 | 33 | 7 | |
| Fisherman (%) | | | | 16 |
| *Secondary* | | | | |
| Goods trader, wage worker, driver, or construction worker (%) | 35 | 31 | 13 | 11 |
| *Spouse* | | | | |
| Homemaker (%) | 40 | 60 | 62 | 44 |
| Store owner, goods trader, wage worker, or ornamental plant seller (%) | 60 | 40 | 38 | 56 |

**Table 11.3** Households' financial status in 2013, based on latest NSCB report of 2009

| | Upstream | Midstream | Upper downstream | Lower downstream |
|---|---|---|---|---|
| Financial status | Pulong Bunga (n=39) | Tartaria (n=39) | Sto. Tomas (n=42) | Aplaya (n=19) |
| Average annual household income | PHP 52,508 | PHP 43,688 | PHP 22,396 | PHP 20,886 |
| 2009 NSCB household poverty threshold | PHP 20,163 | PHP 20,163 | PHP 17,295 | PHP 17,295 |
| % Below poverty threshold | 15 | 21 | 48 | 63 |

PHP 20,163, the poverty incidence in the upstream is estimated at 15 %. Meanwhile, in the midstream areas, with an average annual income of PHP 43,688 for a family of five, poverty incidence is estimated at 21 %. Both these estimates are low compared to the estimated poverty incidences in the downstream sites. In Sto. Tomas, with an average annual income of PHP 22,396 for a family of five, the poverty incidence is estimated at 48 %, while in Aplaya, with an annual income of PHP 20,886, poverty is estimated at 63 % (Table 11.3). These figures do not differ significantly from the figures in the 2009 report by the NSCB, indicating that the economic conditions of fishermen and farmers have been deteriorating. Fishermen posted the highest poverty incidence (41.4 %), followed by farmers (36.7 %).

## *11.4.4 Land-Based Productive Resources*

On average, each household in the upstream *barangay* of Pulong Bunga and the midstream *barangay* of Tartaria is cultivating 1 or 2 parcels of land, respectively.

**Table 11.4** Land-based productive resources: farm holdings, land tenure, and farm distance in Pulong Bunga and Tartaria, Cavite Province, 2013

| | Upstream and midstream *Barangays* | | | |
|---|---|---|---|---|
| | Pulong Bunga (n = 15) | | Tartaria (n = 15) | |
| Total farm area (ha) | No. | % | No. | % |
| ≤1.0 | 11 | 58 | 10 | 67 |
| 1.1–2.0 | 7 | 37 | 5 | 33 |
| 6.0 | 1 | 5 | | |
| Mean | 0.91 | | 1.13 | |
| Range | 0.05–6 | | 0.5–1.5 | |
| Sub-total | 19 | 100 | 15 | 100 |
| Average no. of parcels | 1 | | 2 | |
| *Tenurial status* | | | | |
| Owned | 17 | 90 | 6 | 40 |
| Rented | 1 | 5 | 9 | 60 |
| Rent-free | 1 | 5 | | |
| Sub-total | 19 | 100 | 15 | 100 |
| *Farm distance to* | | | | |
| House | ≤1.0 (14 = 74 %) | | ≤1.0 (10 = 67 %) | |
| Road | ≤1.0 (19 = 100 %) | | ≤1.0 (9 = 60 %) | |
| Market | ≥5.1 (5 = 26 %) | | ≥5.1 (9 = 60 %) | |

The average land area cultivated in Tartaria is bigger, at 1.13 ha, than the 0.91 ha cultivated in Pulong Bunga (Table 11.4). However, 37 % of the farms in Pulong Bunga cover 1.1 to 2.0 ha, in contrast to Tartaria, where the corresponding figure is only 33 %. A majority (90 %) of the farm lots in Pulong Bunga are privately owned, whereas in Tartaria the majority (60 %) are rented. In Pulong Bunga, however, there is also one household cultivating rented land, and another working land rent-free.

Most of the farms cultivated are very accessible to their owners or cultivators; 60–74 % of farms in Pulong Bunga and Tartaria are located less than a kilometer from their owners' or cultivators' homes. The farms are also close to markets, with an average distance of 5.1 km or less.

## *11.4.5 Emerging Issues*

Three major emerging issues are: (1) the incidence of household food insecurity, (2) declining agricultural and fishery productivity, and (3) environmental risks currently affecting food security.

### 11.4.5.1 Incidence of Household Food Insecurity

The incidence of household food insecurity may be understood in the context of the "path to food entitlement," and as such, strategies to address the situation are needed.

The Path to Food Entitlement

Crop losses affect the amount of food available for a family and the amount of food surplus that can be sold for cash to purchase other basic needs. The incidence of food insecurity is therefore influenced by the farming systems adopted (crops grown, number and density of crops, etc.), as well as anthropogenic or biogenic hazards or risks encountered in farm production, and the consequent effect on the family's access to food.

*Difficulties in Meeting Household Expenditures*

Across sites, the majority—if not almost all—of the respondents indicated that they have problems or difficulties in meeting their expenses for even the most important household expenditures. These include food (10–58 %) and farming/fishery inputs (33–69 %). For all sites, the main reasons for problems in paying for their household needs include the seasonal and irregular flow of income, crop losses from typhoons and seasonal climate variations, the high costs of agricultural inputs, and the difficulty of reducing expenses and making other household adjustments. In the upstream and midstream sites, 44 % are below the 2009 NSCB food poverty threshold of PHP 14,040 for the province of Cavite, while 52 % in the upper downstream site and 26 % in the lower downstream site are below the NSCB food poverty threshold of PHP 12,150 for the province of Laguna. Food expenditures include basic necessities such as rice, fish, sugar, coffee, cooking oil, and in some instances canned goods, flavorings, milk, and meat. Rice is bought by the majority of the households across the watershed. Pandey in his report (2008, as cited in Rañola 2010–2011) indicated that in developing countries like the Philippines, poor people spend 30–50 % of their income on rice. However, with the price of rice increasing by as much as 50 %, this means an equivalent drop in their real income of 15–25 %, a substantial loss for poor people who are hovering just above the poverty line.

*Sources of Entitlement, Risks in Farm/Fishery Production, and Seasonal Food Insecurity*

The ability of households and their members to acquire enough food through production, exchange, or transfer depends on their current sources of entitlement. These include their productive assets, such as their land, tools, and equipment, as well as their farm-related entitlements such as the number or density of crops planted, combinations of crop species, period of harvesting and/or production cycle, total land area for cultivation, and the actual farm size cultivated. Other important sources of entitlement include labor resources, education and health, size of the family, level of income, and nature and sources of income. These factors are related to households' levels of poverty and food poverty thresholds as estimated by the NSCB. However, for households in the Silang Santa-Rosa subwatershed, sources of entitlement are limited, and this is compounded by their regular exposure over time to combinations of farm risks and uncertainty (both natural and anthropogenic).

It would therefore appear that they are more or less permanently (chronically) vulnerable, rather than being temporarily (transiently) vulnerable. It becomes apparent that under these conditions, households' paths to food entitlement are being threatened or placed at risk.

#### *Indicators of Food Insecurity and Seasonality of Incidence*

Indicators show that households across the subwatershed are at risk of being food insecure. Many respondents indicated that they had experienced food insecurity to some degree, based on the following indicators of household food insecurity: anxiety about having insufficient food, a food deficit resulting from the absence of funds to purchase food, substitution of cheaper but lower quality food items, and reduction of food intake in terms of quantity and quality. That most of the family income is spent on food is indicative of a food-insecure condition. Across sites, 8–31 % of the respondents constantly experience food insufficiency, while another 44–68 % occasionally have that experience. Approximately 23–100 % of the respondents experience food insufficiency resulting from limited funds to procure food, with the highest percentage of such respondents found in the lower downstream (100 %). More than half of the respondents state that food on the table for the family is inadequate in quantity and quality and that they substitute cheaper foods while reducing the quantity consumed. This is especially the case in downstream areas, being cited by 79 % and 74 % of respondents in Sto. Tomas and Aplaya, respectively. It can be attributed to factors such as the large size of families, low incomes, and the nature of household income sources. Other factors include the number or density of crops planted, crop species and their combinations, harvesting periods and/or production cycles, total land area available for cultivation, and actual farm size cultivated.

### Coping with Food Insecurity

Across sites, the main strategy used by households to cope with food insecurity problems is to generate sufficient cash from non-farm employment to buy their staple food (as cited by 33 % in the upstream, 54 % in the midstream, 57 % in the upper downstream, and 68 % in the lower downstream areas). To address the problem of insufficient funds, about 44–67 % across sites substituted cheaper but lower quality foods than their regular food items, also reducing their food intake. Farmers in the upstream and midstream, moreover, try to produce nearly all of their own staple food requirements like vegetables and fruit crops. It is apparent, however, that although the households have instituted some measures to address their conditions, these measures reflect the limitations in farm households' assets. For the most part, these adaptive measures are ad hoc and limited in mitigating the impacts of low farm income resulting from the biophysical risks, both natural and man-made, that are inherent in farming.

#### 11.4.5.2 Declining Agricultural and Fishery Production

Farm households across sites are faced with declining agricultural production. This is attributable to the biophysical traits of their farms, as well as risks associated with typhoons, strong winds, El Niño, and other seasonal climate variations. Socio-economic and institutional factors have also played a part. The reduced agricultural productivity in farmlands has adversely affected the level of households' farming income. In the upstream, 40 % of farmer respondents stated that their income had declined over the 5 years from 2008 to 2012, with their usual average income of PHP 180,334 dropping by as much as 42 % over the 5-year period. In the mid-stream, 20 % experienced decreases in income by as much as 58 %, in the upper downstream, 50 % cited a decline in production and consequent 50 % decrease in income. In the lower downstream, about 33 % of fishermen in Aplaya experienced a reduction in household income due to diminished fish catches. Their income dropped by as much as 70 % from the usual average annual income of PHP 160,000. Other factors contributing to the decline in fishery income are natural factors such as typhoons, heavy rains combined with strong winds and tidal waves, and polluted water conditions in the lake that contribute to the poor growth of fish.

#### 11.4.5.3 Environmental Risks Currently Affecting Food Security

Farmers and fishermen perceive the risks associated with environmental degradation and its long-term negative consequences as affecting not only themselves but the community as a whole. The risks are of both natural and anthropogenic origins. Those of natural origin include typhoons, strong winds, El Niño, and seasonal variations brought about by climate and/or weather disturbances. In Silang-Santa Rosa, these have the effects of altering the crop cycle, and reducing crop productivity and income. Other types of risks include outbreaks of pests and diseases, crop losses from toxic chemicals, soil erosion and acidity, depressed market prices, and land use conversion.

Farmers and fishermen have faced these risks to some degree for years. However, in the upstream, 46 % claimed that their level of exposure to such risks is low. A low level of exposure was also cited by 25 % in the midstream, 100 % in the upper downstream, and 50 % in the lower downstream. This can be attributed to the respondents' resiliency and the measures they have adopted to address the risks. Factors affecting their level of resiliency include their earning capacity, family assets, and asset management, as well as support from family and relatives that provide financial assistance in times of need. One specific adaptive measure taken in Silang-Santa Rosa is to immediately replant to replace damaged crops. Otherwise farmers simply wait for crops to recover naturally, or go and seek off-farm or non-farm employment.

However, these adaptive measures are generally makeshift and do little to reduce the impacts of climatic events and other forms of risks. They are also very dependent on the level of household assets and resources. According to the respondents, in the longer term these risks will pose a major threat to the income, food security, and health of their families and community. They also mentioned that with a persistently low farm productivity and income, they would be forced to stop sending their children to school, or possibly have no choice but to sell their farmland. Inadequate food supply would also increase the incidence of hunger, malnutrition, and possibly even death. For the community as a whole, it would lead to loss of their crops and farms, and the destruction of their properties.

## 11.5 Conclusion

The incidence of food insecurity can be viewed as the ultimate effect of several inter-related biophysical and socio-economic factors that comprise the main hazards and risks. Environmental degradation, for example, may be caused by a combination of man's interaction with physical resources and naturally occurring hazards. It is evident in all the sites that the declining land resource base is adversely affecting farm and fishery households' incomes and livelihoods, resulting in the threat of food insufficiency for the poorest. Across sites, households perceive that continued occurrence of climatic and other natural events in the years to come will have a major impact on their incomes, family health, food security, and the circumstances of the community as a whole.

With these issues in mind, the following technical and policy concerns should be taken into consideration to develop a research agenda that would facilitate the creation of a plan for managing watersheds and implementing related strategies.

1. Developing an integrated approach to watershed management that covers both land and water resources entails convening the major stakeholders to: (1) develop and agree on a common vision, goals, and objectives for managing the watershed and lake resources (e.g., improved water quality for a given period of time); (2) identify issues and problems that may hinder the fulfilment of the vision, goals, and objectives; and (3) prioritize and develop an agreed action plan for dealing with the identified problems and/or constraints.
2. In trying to address issues relating to soil and land degradation, it is important to determine whether the causes are anthropogenic or geogenic, since the potential solutions would be different. For example, many problems related to soil and land degradation can be traced to watershed management practices that are dictated by stakeholders' socio-economic circumstances. Of research interest, therefore, are studies on the appropriateness of different technologies given different physical and biological conditions of the soil. Technologies that address different bio-physical and socio-economic conditions among farmers might include re-vegetation, erosion control measures, water conservation, agroforestry systems, organic farming, and use of organic pesticides.
3. When developing plans for the watershed areas, addressing poverty and providing the means to make a living at the household level should be a major consideration, given our understanding of the relationships between land and water resources

and the impacts of land use on water resources. Policymakers should understand, for example, that farming and fishing are very important to households since they provide a steady source of income despite their relatively small contribution to the family finances overall. It is therefore important to promote farm production and fishing practices that not only protect natural and environmental resources, but also improve households' levels of income. In addition, there should also be occupational opportunities for those children with better education who may decide to opt out of farming or fishing.

4. There is a need to assess household food risk as well as other forms of risk associated with the livelihoods of subwatershed residents. Households perceive that continued climatic events and other natural phenomena in the years to come will have a major impact on their incomes, family health, food security, and the community as a whole. Moreover, both the current study and previous studies have indicated that the incidence of food insecurity at the household level is influenced by such factors as the seasonality of farm income. This seasonality is in turn affected by the types of cropping systems employed and crops grown, the size of tilled land, poor harvests due to variable climate conditions, price volatility, and low food stocks.
5. Soil and water pollution, soil erosion and degradation, and adverse effects on health are also major issues faced by the communities in the study. Important questions that need to be answered are: What are the economic costs of soil erosion or degradation? What are the economic costs of ill health or sickness due to pollution? What economic and regulatory instruments can be instituted to reduce the level of soil and water pollution, soil erosion, and degradation? Determining the economic value of potential losses and gains from different measures is important for making policy and investment decisions to address environmental degradation.



## References

Food and Agriculture Organization-Food and Nutrition Division, Publishing Management Group (1997) Agriculture food and nutrition for Africa: a resource book for teachers of agriculture, p 116. http://www.fao.org/docrep/w0078e/w0078e00.HTM. Accessed 25 Aug 2013

National Statistical Coordination Board (2009) Philippines policy making and coordinating body on statistical matters. Philippine poverty statistics. http://www.nscb.org.ph

Rañola RF Jr, Rañola FM, Casin CS, Tan MFO (2010–2011) LakeHEAD progress report: the social and economic basis for managing environmental risk for sustainable food and health in watershed planning: the case of Silang-Sta. Rosa sub-watershed communities in Lake Laguna region. Research Institute for Humanity and Nature, Japan

Springer Editorial (2009) Food security and global environmental change: emerging challenges. Elsevier Environmental Science and Policy 12:373–377. http://www.elsevier.com/lacate/envsci. Accessed 23 Aug 2013

# Chapter 12
# Strengthening the Capacity of Flood-Affected Rural Communities in Padang Terap, State of Kedah, Malaysia

**Fera Fizani Ahmad Fizri, Asyirah Abdul Rahim, Suzyrman Sibly, Kanayathu C. Koshy, and Norizan Md Nor**

**Abstract** The communities in the district of Padang Terap, Kedah, were not used to flood events even though they had been living in the area for generations. Since 2000, flooding in this area had become a common occurrence as a result of the increased intensity and frequency of rain due to global warming and climate change. Recognizing the need to address the vulnerability and adaptation of the communities and relevant agencies, the Centre for Global Sustainability Studies (CGSS), Universiti Sains Malaysia (USM), carried out a project based on university-community engagement and education for sustainable development (ESD) in this area. The stakeholders involved in this project were from different entities, consisting of experts from USM, residents, local officials, and selected schools in the district. The project identified that the communities of Padang Terap needed counseling for trauma victims, training in handling and maintaining flood-related equipment, accredited flood rescue training, and the establishment of a formal community flood disaster committee consisting of village leaders and the local authorities. In addition, CGSS also addressed public awareness of the dangers of flooding via ESD

F.F.A. Fizri (✉)
School of Industrial Technology, Universiti Sains Malaysia, 11800 USM, Penang, Malaysia
e-mail: fera@usm.my

A.A. Rahim
School of Humanities, Universiti Sains Malaysia, 11800 USM, Penang, Malaysia
e-mail: asyirah@usm.my

S. Sibly • K.C. Koshy
Centre for Global Sustainability Studies, Universiti Sains Malaysia,
11800 USM, Penang, Malaysia
e-mail: suzyrman@usm.my; koshy_k@usm.my

N.M. Nor
School of Humanities, Universiti Sains Malaysia, 11800 USM, Penang, Malaysia
e-mail: norizan_mdnr@yahoo.com

N. Kaneko et al. (eds.), *Sustainable Living with Environmental Risks*,
DOI 10.1007/978-4-431-54804-1_12, © The Author(s) 2014

activities implemented in schools and villages. Close cooperation and positive contributions from academia, local officials, and local communities ensured that the project was successful.

**Keywords** Capacity building • Community engagement • Disaster preparedness

## 12.1 Introduction

Flooding is a natural disaster caused by climatological factors such as temperature, rainfall distribution, evaporation, wind movements, and the natural terrain (Balek 1983). Fauchereau et al. (2003) and Camerlengo and Somchit (2000) have attributed the change in rainfall distribution to the global warming phenomenon. Flooding can be categorized into river flooding, flash floods, and storm surges. River flooding is caused by heavy and/or continuous rainfall over a period of a few days or weeks in a large area. An important characteristic of this type of flooding is the soil, which becomes saturated, exceeding its capacity to absorb water, and thereby increasing overland flow and water retention (Kron 2002; Berz et al. 2001).

In Malaysia, the flood disasters that occur are due to flash floods, tropical storms, and monsoon storms. On the east coast of peninsular Malaysia, particularly in the states of Terengganu, Kelantan, and Pahang, flooding normally occurs in the rainy season and the frequency of these floods is affected by changes in the monsoon seasons (Chan 1996; Jamaluddin and Sham 1987; Rose and Peter 2001). During the monsoon seasons there is a continuous and increased amount of rainfall, causing the river waters to overflow.

Climate change is seen as a global phenomenon; however, its impacts are localized and long-term. The effects of climate change are evident in the increased occurrence of flooding in the coastal areas of Kedah, Kelantan, Terengganu, Pahang, and Johor. Continuation of this scenario will have a significant impact on the society's culture and economically sensitive sectors as well as on the well-being of those affected.

Among the impacts of floods are pollution; erosion; damage to building structures; loss of property; loss of life; damage to the drainage system; contamination of food and water; disruption of socio-economic activities, including the transportation, telecommunications, and services network; and loss of environmental services resulting from effects such as the degradation of agricultural soil. Flood occurrences in 2006 and 2007 amounted to RM 1.1 billion and RM 776 million in losses, respectively. This amount reflects only the losses incurred by the Malaysian government, and does not take into consideration losses sustained by flood victims and by the local economy.

Those worst affected by the floods are residents who are weak and less well prepared. One of the strategies for flood risk management, therefore, is to make those vulnerable to flood disasters more resilient and better prepared. Efforts need to be focused toward mitigation and enhancing the adaptability or coping capabilities of these residents.

## 12.2 Theoretical Framework

The Hyogo Framework for Action 2005–2015 document calls on us to promote a strategic and systematic approach to reducing vulnerabilities and susceptibility to hazards. According to the United Nations Office for Disaster Risk Reduction, vulnerability is defined as the conditions determined by physical, social, economic, and environmental factors or processes that increase the susceptibility of a community to the impact of hazards (United Nations International Strategy for Disaster Reduction (UNISDR) 2004a). Viewed in combination with risk, hazards, and capacity, the following equation is derived (United Nations International Strategy for Disaster Reduction (UNISDR) 2004b):

$$\text{Risk} = \text{Hazards} \times \text{Vulnerability} / \text{Capacity}$$

Looking at the equation above, it is logical to say that by increasing capacity and reducing vulnerability, risk can be minimized. The Hyogo Framework for Action (United Nations International Strategy for Disaster Reduction (UNISDR) 2007) listed five priorities for action:

1. Ensure that disaster risk reduction is a national and a local priority with a strong institutional basis for implementation
2. Identify, assess and monitor disaster risks and enhance early warning
3. Use knowledge, innovation and education to build a culture of safety and resilience at all levels
4. Reduce the underlying risk factors
5. Strengthen disaster preparedness for effective response at all levels

Using this and the above equation as a basis, capacity building initiatives should be at the forefront of disaster risk management activities.

## 12.3 Case Study

### *12.3.1 Study Area*

The area chosen for this study focuses on the district of Padang Terap, Kedah, based on a situational analysis conducted on rain profiling. Padang Terap district was the area most affected during flood disasters from 2000 until 2010. The area covers 135,684.41 ha and the population size is 72,318, consisting of Malay and Malay-Thai (DID 2010). Padang Terap district has 12 *mukims* (sub-districts), and one village head is appointed for each *mukim*. The study area is shown in Fig. 12.1.

The district of Padang Terap is the second largest district in the state of Kedah. The main economic activities are rubber tapping and farming, and the principal crop grown is paddy. The district's farmers do not, however, plant paddy on a large scale

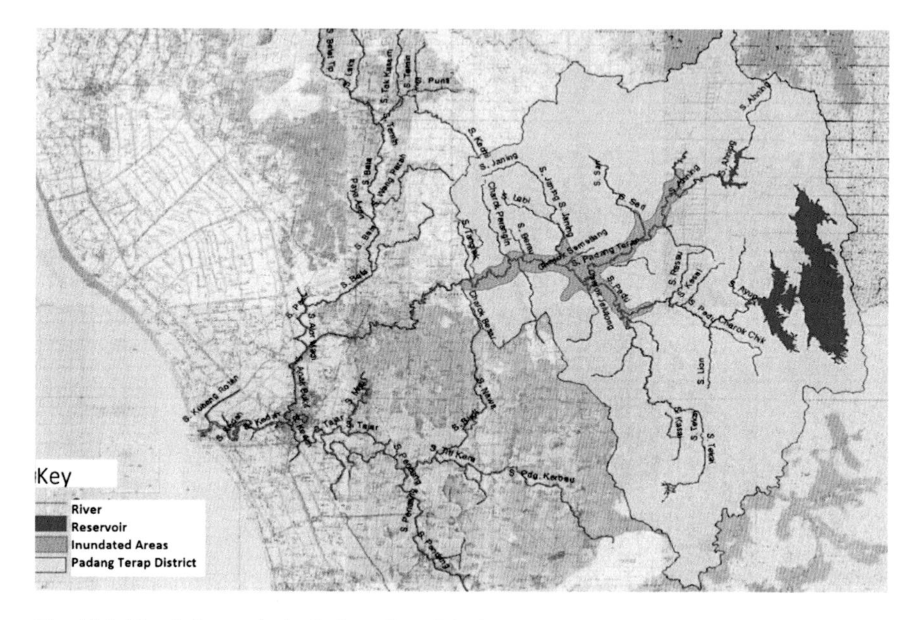


**Fig. 12.1** Flooded areas in the Padang Terap District

and are highly dependent on the river and rain water for irrigation. They are very much exposed to the threat of losing their crops and livelihoods if and when a flood occurs in the area.

Flooding occurrences have increased in Padang Terap, happening at least once, and sometimes up to eight times, in a year. The duration of the floods averages about three days, but can last up to fourteen days consecutively.

## *12.3.2 Needs Analysis*

In order to curb losses and utilize resources in a more efficient manner, this study focuses on identifying community needs during floods in the district of Padang Terap, Kedah. A survey of the entire population of flood-affected areas was conducted to gain an insight into population demographics and flood related losses. Furthermore, focus group discussions with village heads were held to identify common issues and needs.

### 12.3.2.1 Survey

A survey was conducted to assess the flood victims' needs during flood disasters, as well as to develop a valid and verified set of baseline data. The sampling method

**Table 12.1** Respondents' needs during floods

| Needs | Number of respondents |
|---|---|
| Assistance to move and lift belongings | 146 |
| Food supplies | 96 |
| Access to communication | 88 |
| Rescue assistance | 45 |
| Assistance in moving to a safe place | 31 |
| Access to electricity, water, and other utilities | 17 |
| Quicker/faster flood assistance | 14 |
| Assistance with watching over assets | 10 |
| Medical care | 7 |

**Table 12.2** Respondents' needs after floods

| Needs | Number of respondents |
|---|---|
| Assistance to tidy and clean up | 579 |
| Assistance to replace damaged belongings | 33 |
| Medical care | 8 |
| Food supplies | 8 |
| Monetary assistance | 3 |

chosen for this study was purposive sampling, whereby the respondents selected were those directly involved in paddy farming, vegetable farming, and other types of agricultural activity.

Based on the existing data from the district office, 62 *kampungs* (villages) under 11 *mukims* were identified as flood-affected areas in the Padang Terap district. Once the *kampungs* and *mukims* associated with floods were identified, specific flood areas were located with help from *ketua kampung* (village leaders). The survey was then carried out in the specifically identified flood areas.

The survey was distributed to 683 flood victims to obtain their demographic information, to identify vulnerable groups and affected areas, and to identify the needs of the victims before, during, and after a flood occurs. The data analysis was conducted using SPSS Software version 17.

The total number of flood victims in the 11 *mukims* was 683. The *mukims* of Belimbing Kiri and Belimbing Kanan had the highest number of flood victims at 179 and 172, respectively, accounting for 51 % of the total flood victims in Padang Terap.

The educational levels of the flood victims were found to be quite low. The majority of respondents (538 respondents, 79 %) had experienced primary and secondary education only, while 99 respondents (14.5 %) had no formal education.

As shown in Table 12.1, during floods the majority of respondents needed one or more of the following: (1) assistance with moving and lifting belongings, (2) food supplies, (3) access to means of communication, (4) rescue assistance, or (5) assistance in moving to a safe place. After flooding, 579 respondents required assistance to tidy up and clean their houses and equipment such as high-pressure jet sprays (Table 12.2).

Table 12.3 Respondents' actions before, during, and after floods

| | Number of respondents |
|---|---|
| *Before floods* | |
| Did not save contingency money | 626 |
| Elevated belongings | 145 |
| Safeguarded important documents | 101 |
| Moved vehicles to higher ground | 97 |
| Bought food supplies | 84 |
| Saved contingency money | 57 |
| Moved to a safer place | 19 |
| Built higher/elevated home | 3 |
| *During floods* | |
| Moved to flood shelter | 228 |
| Moved belongings to a safer place | 184 |
| Contacted relevant authorities | 73 |
| *After floods* | |
| Cleaned mud from house | 440 |
| Attended medical checkup | 127 |
| Disinfected belongings, etc. | 104 |
| Sought counseling | 15 |

When asked about their preparations before floods occur, only a handful of the respondents had made monetary preparations for flooding (57 respondents, 8 %). Most of the respondents did not make any monetary preparations, and despite floods being a common occurrence since the year 2000, only three respondents (0.4 %) had adapted by building elevated homes (Table 12.3).

Most of the respondents appeared to be aware of what needs to be done during flooding, with 228 (33 %) stating that they would move to flood shelters, and 184 (27 %) saying that they would move their belongings to a safer place (Table 12.3).

After floods subsided, most of the respondents (440, 64 %) concentrated their efforts on cleaning the flood residue from their houses. Another 104 respondents (15 %) started making arrangements to disinfect their houses, and 127 respondents (18 %) went for a post-flood medical checkup (Table 12.3).

#### 12.3.2.2 Focus Group Discussions

A qualitative approach, the Focus Group Discussion (FGD), was used to identify problems and needs from the flood victims' perspective. This method was used to obtain a better understanding of specific issues, which is vital as it can lead to implementation of appropriate and effective solutions before, during, and after floods. The main stakeholders involved in the FGD were representatives from the *Majlis Ketua Kampung* (Village Head Council) of Padang Terap and the *Jawatankuasa Kemajuan & Keselamatan Kampung* (Village Development & Security Committee).

Since there are two different village-level administrative committees representing the federal government and state government, respectively, two separate FGD groups were convened. One catered to the federal government village committee, the other to the state government village committee. A total of 32 representatives from 29 *kampungs* (villages) attended the FGD sessions. Of the 32 representatives, 18 were representatives of the state committee while the remaining 14 were representatives of the federal government committee.

The FGD primarily focused on the problems faced by both the federal and state representative committees related to flood occurrences. The purpose of this exercise was to provide a basic understanding of the major issues faced by the authorities in particular before, during, and after a flood occurs. Among the major issues raised during the FGD were:

1. Dissemination of information on rising flood water levels to flood affected communities was slow and inefficient.
2. Victims in flood shelters did not receive food supplies on time due to delays in the supplies reaching the shelters.
3. There were insufficient rescue boats available for mobilization during flooding.
4. Other equipment, such as high-pressure water jet sprays to clean houses after flooding, tents for constructing emergency shelters, and portable gas-powered generators to provide electricity were lacking, despite being urgently required to prepare for future flood occurrences.

### 12.3.3 Capacity Building Activities

Several measures were identified to strengthen capacity within the local community and Padang Terap's local authority. Bearing in mind that children and the elderly were the most vulnerable, a *Bahaya Ayaq Bah* awareness campaign was conducted in six schools within the Padang Terap district. *Bahaya Ayaq Bah* is a phrase in the local dialect meaning "The Dangers of Flooding." The schools comprised Sekolah Kebangsaan Toh Puan Syarifah Hanafiah, Sekolah Kebangsaan Kurong Hitam, Sekolah Kebangsaan Padang Sanai, Sekolah Kebangsaan Kubang Palas, Sekolah Kebangsaan Kuala Nerang, and Sekolah Kebangsaan Seri Bakti. These schools are among those that, in the event of a flood, would be either inundated, surrounded by flood waters, or turned into disaster relief centers. In addition to being educated on the dangers of flooding, pupils were also exposed to what needs to be done before, during, and after a flood occurs. The campaign comprised activities in the form of crossword puzzles, quizzes, and short talks on flooding, as well as an introduction to the 3S concept.

The 3S concept (*Sebelum, Semasa, Selepas*) was coined from the local terminology for "before," "during," and "after" flood occurrence. Information relating to the three stages was disseminated to the students, advising them to ascertain if their homes were vulnerable to flood, to evacuate their homes once the flood warning was issued by the authorities, and to ensure the whereabouts of each family member during the evacuation process. In addition, the students were reminded to be careful about personal protection, for example by wearing gloves during the cleaning process, and to clean and disinfect furniture and appliances thoroughly to avoid waterborne or water-related diseases.

Coping capacities are defined as the ability of a society or group, organization, or system to use its own resources to address and manage emergencies, disasters, or adverse conditions that could lead to a harmful process caused by a hazard event (UNISDR 2009). To increase the coping capacities of the local community, residents were introduced to the concept of a "flood kit." According to the survey conducted, the local community experienced difficulty with evacuating or being rescued. Due to the high number of victims and insufficient resources, evacuation was subject to delays. In order for the locals to cope with the situation, they were encouraged to prepare a flood kit. This would consist of basic first aid items and toiletries, a flashlight, a bottle of water, food items such as instant noodles and biscuits, and a plastic folder to hold important documents. The flashlight would be useful during the night, especially if the electricity had been cut off, and the food items would tide the family over until help arrived. Every household was encouraged to prepare a flood kit for each family member.

In addition to preparing families for disaster, training was also given to the local community on handling and maintenance of rescue boats. Incorporating rescue measures and procedures, and proper usage of life vests and floatation devices, the training was conducted by the Public Defense Department, the Fire and Rescue Department, the Police Department, and the Muda Agriculture Development Authority (MADA).

## 12.4 Future Research and Recommendations

This project covered only certain issues identified in the survey and raised during the FGD. Other issues such as insufficient communication and long-term adaptation strategies will need to be addressed in future. Implementation of an early warning system or a flood alert is also important. Warning systems will need to be efficient to enable proper preparation for evacuation. These systems have to run without electricity and reach even the most secluded homes.

**Acknowledgment** We would also like to express our utmost gratitude to this project's research officer, Mr. Mohd Zulhafiz Said, and his assistants employed under the USM APEX Delivering Excellence grant, for collecting data for this project.

## References

Balek J (1983) Hydrology and water resources in tropical regions. Dev Water Sci 18:216–235

Berz G, Kron W, Loster T, Rauch E, Schimetschek J, Schmieder J, Siebert A, Smolka A, Wirtz A (2001) World map of natural hazards–a global view of the distribution and intensity of significant exposures. Nat Hazards 23:443–465

Camerlengo AL, Somchit N (2000) Monthly and annual rainfall variability in peninsular Malaysia. Pertanika J Sci Technol 8(1):73–83

Chan NW (1996) Vulnerability of urban areas to floods. Star 1996:4–6. Climate Change Research, UK

Department of Irrigation and Drainage (2010) Kedah flood report. Department of Irrigation and Drainage, Kedah

Fauchereau NS, Trzaska MR, Richard Y (2003) Rainfall variability and changes in Southern Africa during the 20th century in the global warming context. Nat Hazards 29:139–154

Jamaluddin J, Sham S (1987) Development process, soil erosion and flashfloods in the Kelang Valley Region, Peninsular Malaysia: a general consideration. Arch Hydrobiol Beih 28: 399–405

Kron W (2002) Flood risk = hazard × exposure × vulnerability. In: Wu B, Wang ZY, Wang G, Huang GGH, Fang H, Huang J (eds) Flood defense. Science Press, New York

Rose S, Peter EN (2001) Effect of urbanization on stream flow in the Atlanta area (Georgia, USA): a comparative hydrological approach. Hydrol Process 15(8):1441–1457

United Nations International Strategy for Disaster Reduction (UNISDR) (2004a) Living with risk: a global review of disaster reduction initiatives. Preliminary version. Inter-Agency Secretariat of the International Strategy for Disaster Reduction, Geneva, p 382 http://www.unisdr.org/eng/about_isdr/bd-lwr-2004-eng.htm

United Nations International Strategy for Disaster Reduction (UNISDR) (2004b) Disaster risk reduction tools and methods for climate change adaptation. UNISDR, Geneva

United Nations International Strategy for Disaster Reduction (UNISDR) (2007) Hyogo framework for action 2005–2015: building the resilience of nations and communities to disasters. United Nations, Geneva

United Nations International Strategy for Disaster Reduction (UNISDR) (2009) Global assessment report on disaster risk reduction. Risk and poverty in a changing climate – invest today for a safer tomorrow. United Nations, Geneva

# Chapter 13
# Mitigating Coastal Erosion in Fort Dauphin, Madagascar

**Jean-Jacques Rahobisoa, Voahangy Rambolamanana Ratrimo, and Alfred Ranaivoarisoa**

**Abstract** The city of Fort Dauphin is one of the most attractive tourist spots in Madagascar. In recent years, it has become one of the development centers of the island. However, the city is facing coastal erosion related to human activity as well as natural factors. Mitigation of coastal erosion at the catchment and regional scale is extremely important for the sustainable economic and social development of this region. Spatial analysis using satellite imagery over a long period has been considered as an important tool for determining the extent of the most affected areas and for analyzing how the erosion has developed in the past and at present. Measurement and field work need to be integrated to develop appropriate strategies to mitigate the problems. Satellite imagery analysis in combination with field work and measurement consists of generating long-term information required to determine threats and pressures in time and space. It takes into consideration assessment of land use, the geology of the area, urban planning, local and regional climate, and coastal management. Madagascar faces multiple challenges in mitigating coastal erosion, but the involvement of authorities and local communities plays a key role in long-term shoreline protection.

**Keywords** Fort Dauphin • Mitigation of coastal erosion • Spatial analysis

## 13.1 Introduction

Due to its growing intensity, coastal erosion seems to be an abnormal phenomenon, but it is in fact a common occurrence. It has served as the key factor shaping coastal environments throughout history (Niesing 2005; Prasetya and Black 2003). Coastal systems play a variety of roles including assimilation of wave energy, hatching of

J.-J. Rahobisoa (✉) • V. Rambolamanana Ratrimo • A. Ranaivoarisoa
Department of Earth Sciences, Faculty of Sciences, University of Antananarivo, Post Box 906, Antananarivo, Madagascar
e-mail: rahobisoajacques@yahoo.fr; ratrimovoahangy@yahoo.fr; alfredranaivo@yahoo.com

N. Kaneko et al. (eds.), *Sustainable Living with Environmental Risks*,
DOI 10.1007/978-4-431-54804-1_13, 

flora and fauna, and groundwater protection, and they contribute significantly to recreational activities (Niesing 2005). Fort Dauphin is among the coastal zones threatened by coastal erosion. A recent rise in sea level, landslides, and coastal erosion have become serious threats to this municipality. Since 2005, coastal erosion has resulted in loss of housing facilities, four recreational beaches, and road communication links. In 2011 and 2012, widespread diminishing of beaches was observed around the city, with some areas badly damaged. Coastal erosion can be classified as a major risk for the city due to the threat it poses to economic development activities. In light of this, in 2012 the government decided to finance coastal erosion surveys through the PIC (Pole Integré de Croissance) project.

According to the Intergovernmental Panel on Climate Change (IPCC) in its Fourth Assessment Report in 2007, climate change marked by increasing temperature and a rise in sea level will place populations living in coastal areas in grave danger in 80 years' time. Moreover, Key human activities (Neuvy 1981; Schiereck 2004; Williams and Micallef 2009; Brebbia et al. 2009; Kim 2010; Slovinsky 2011) are located in coastal zones. In other words, the pressures of climate change and increase in human activity toward the coasts has turned coastal erosion into a more serious problem, not only for coastal municipalities, but for the world as a whole (Prasetya and Black 2003; Niesing 2005). Coastal erosion is usually the outcome of many factors in combination including natural and human-induced influences operating on different scales (Shore Protection Manual 1984; Ir Zamani Bin Mindu 1988; DGENV European Commission 2004; Williams and Micallef 2009; Slovinsky 2011). In undertaking this case study, therefore, it was necessary to understand the factors responsible for coastal erosion in order to develop mitigation strategies.

Fort Dauphin is an urban municipality in the southern part of Madagascar. According to the administration division, it is the capital of Anosy Region, and Amboasary district is positioned at the south eastern limit of Madagascar, in an area where winds blow from the Indian Ocean (Fig. 13.1). Geomorphologically, this study area is characterized in its eastern part by a small coastal plain dominated by the Anosyan Mountains. Morphological disposition makes the region highly exposed to winds blowing from the east (alize) and orographic lifting occurs frequently (Ratsivalaka Randriamanga 1985). The contacts with the ocean in the eastern and southern parts of the study area constitute cliffs varying between 5 and 10 m in height.

### 13.1.1 Climate

Under the influence of the Indian Ocean and its morphology, this region receives significant rainfall; its average annual rainfall is 1,800 mm/year (Direction de la météorologie 1980). As in the other regions in the country, December and January are the rainy months, comprising the rainy season. As it is exposed to the Indian Ocean, however, rainy and dry seasons are sometimes mixed. Temperature in this area varies between 20 and 26 °C (Direction de la météorologie 1980), where the maximum temperature corresponds to the rainy season and, conversely, the minimum temperature typifies the dry season.

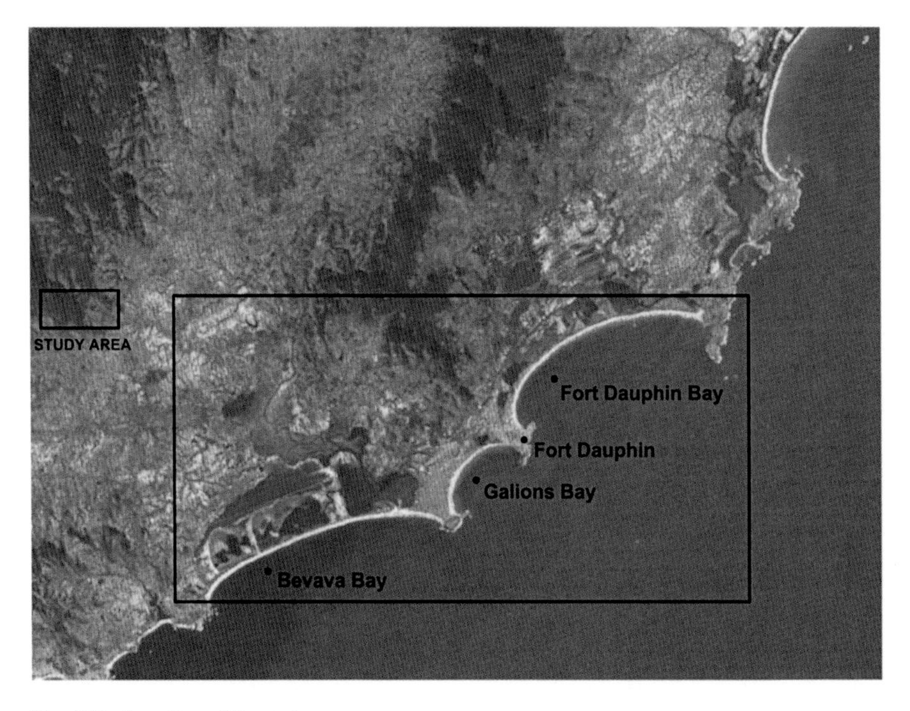


**Fig. 13.1** Location of the study area

### *13.1.2 Geology*

Fort Dauphin essentially comprises sedimentary rocks that form Pleistocene and Holocene dunes. The eastern and southern areas are characterized by cemented dunes, where limestone playing the role of cement makes relatively hard rock. Due to its hardness, it forms cliffs in some areas. Westward, this formation is topped by unconsolidated recent dune, which occupies three fourths of the area. In the north east of this locality, metamorphic rocks of the Paleoproterozoic Era (1.8 Ga) act as a climatic barrier (Fig. 13.2).

## 13.2 Methodology

The development of Fort Dauphin's coastal degradation was studied by comparing satellite imagery taken at different times, namely 1998, 2004, 2009, and 2011. These satellite images enabled us to measure and observe the variation of the beaches and the coastal area width (Cambers 1998; Brebbia et al. 2009; Y Wang 2010). In this survey, the most damaged areas in each of the years were analyzed, such as Bevava Beach, Galions Bay, and Fort Dauphin Bay, as well as Libanona

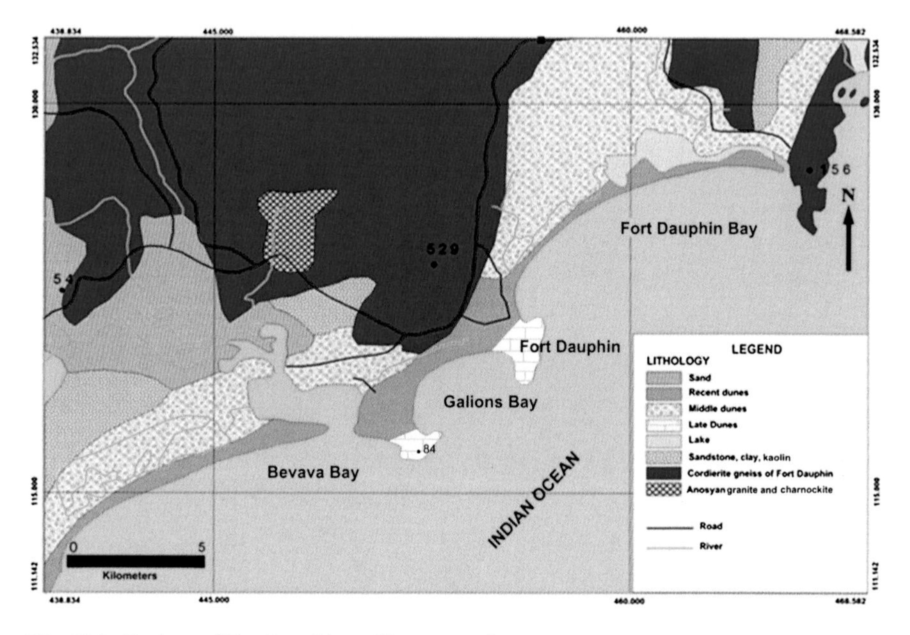


**Fig. 13.2** Geology of Fort Dauphin and its surrounding areas

Beach. The dunes of Galions Bay were also studied due to the combination of coastal and aerial erosion. The analysis involved measuring a fixed 20 m portion of the bays for each of these periods; 1998 was taken as the reference point in these measurements, and the analyses incorporated some measurements of water levels, waves, and currents made by QIT Madagascar Minerals (QMM).

These cartographical analyses were coupled with field work and villagers' observations, where coastal erosion was observed for each measurement at different periods. Some photos and video were obtained from the fieldwork, facilitating the formulation of appropriate solutions to stop or mitigate the coastal erosion Phenomenon (Auckland Regional Council 2000; DGENV European Commission 2004; Brebbia et al. 2009; NSW Government 2010; Y Wang 2010).

### *13.2.1 Data Processing*

Numerical models were also used to assess and analyze the waves in the surrounding area and around Ehoala Port. The results of these models were checked using satellite, recorded wave, and pressure data. Based on satellite imagery analyses, Orthophoto, Quickbird, and worldview respectively for 1998, 2004, 2009, and 2011 were treated with GIS assessment methods using Arc gis.

## 13.3 Results and Discussion

Table 13.1 summarizes the development of coastal erosion in Fort Dauphin, which manifests by reduction of beach width in the four selected portions that are most affected by coastal erosion (Fig. 13.3). In some periods, the phenomenon of accretion was also apparent, but in general, diminishing of the beach width ranged from 2.6 m (Galions bay) to 9 m (Ambinanibe). The accretion measured during the study periods was insignificant compared to the sediment loss.

**Table 13.1** Development of coastal erosion in Fort Dauphin

| Period | Ambinanibe Beach (630 m) | Galions Bay (400 m) | Libanona Beach (430 m) | Fort Dauphin Bay (1,000 m) |
|---|---|---|---|---|
| 2004–2009 | Accretion = 9.4 cm/year | Loss = 0.87 m/year | Accretion = 29 cm/year | Accretion = 30.4 cm/year |
| 2009–2011 | Loss = 9.165 m/year | Loss = 2.64 m/year | Loss = 4.81 m/year | Loss = 5.265 m/year |

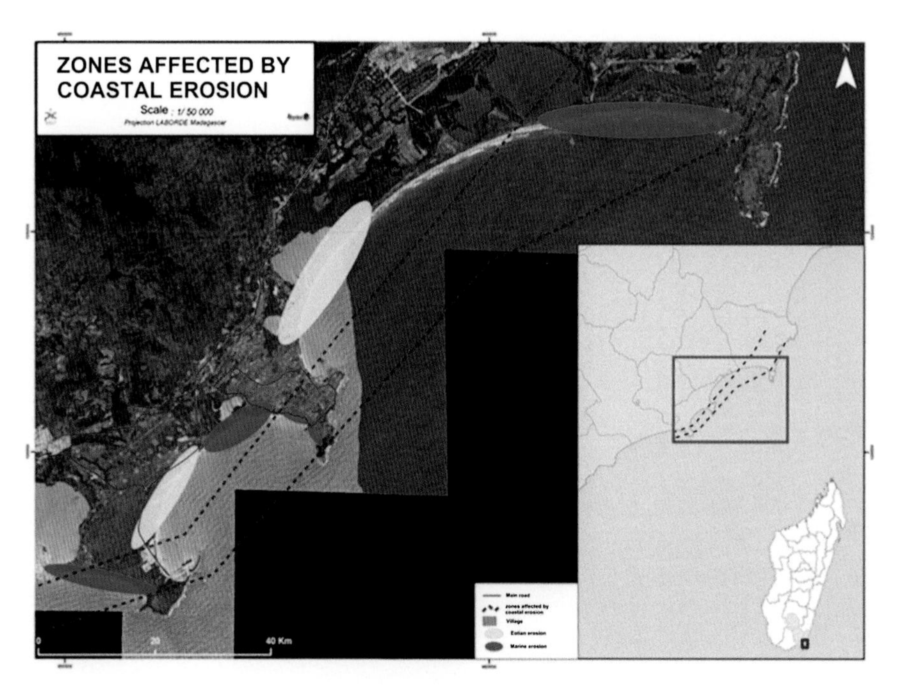


**Fig. 13.3** Selected portions showing coastal erosion and wind erosion respectively in *blue* and *yellow*

### 13.3.1 Impacts and Causes of Coastal Erosion

The impacts of coastal erosion most frequently encountered in Fort Dauphin can be grouped into three categories: coastal flooding as a result of dune erosion, undermining of sea defenses associated with foreshore and subaerial erosion, and retreating cliffs, beaches, and dunes causing loss of land (Fig. 13.4).

Coastal erosion is derived from numerous causes but wind and current are particularly significant. These two parameters play important roles in coastal abrasion. Due to the similarity in the locations of the most affected areas, and the wind

**Fig. 13.4** Impacts of coastal erosion in Fort Dauphin

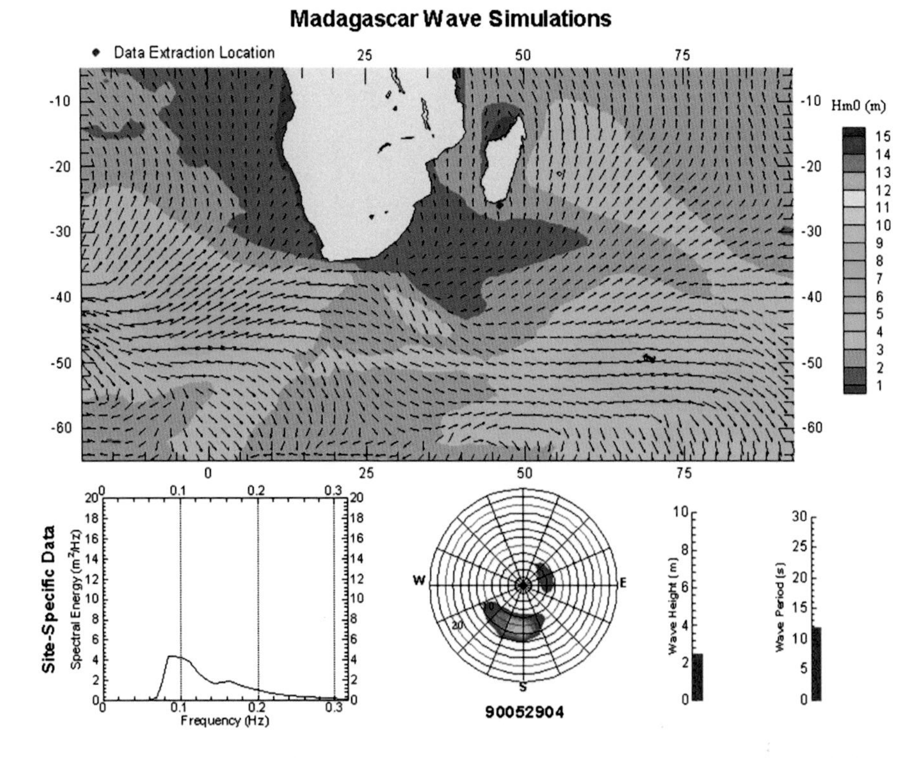


**Fig. 13.5** Wave simulations in Fort Dauphin showing wave directions (Rio Tinto-QMM 2008)

and current directions, the results of numerical models of waves and current direction show that wave action and ocean currents are among the most important factors causing coastal erosion in Fort Dauphin. As shown in Figs. 13.5, 13.6, 13.7, and 13.8, the majority of the strongest waves and currents come from the southwest and east.

The manifestation and activity of these two parameters on beach and dunes could be explained by combinations of various natural forces such as the wave direction approach, as well as the dredging (digging) phenomenon in the coastal area. Before explaining the coastal erosion process in Fort Dauphin, it is interesting to remember that the surf zone is the area where waves break. It is a turbulent zone, as waves smash and dissipate their energy in this area while producing intense local currents that eventually reach the coastal shores (Hyndman 2006). During this turbulent time, water removes sediment in its path and then local currents carry it to the sea when leaving the coast. The same process occurs in the wave zone; in this zone, the depth of breaking varies depending on wave size (Hyndman 2006). This phenomenon is observed on all the beaches in Fort Dauphin, and beach and dune dredging appears in the same manner as shown in Fig. 13.9.

During the study, natural sand transport into the deep sea was observed in almost all the beaches of the city after dredging. Sand content in sea water varies depending on the area, but it seemed greater in Bevava Bay (Ambinanibe) and Galions Bay.

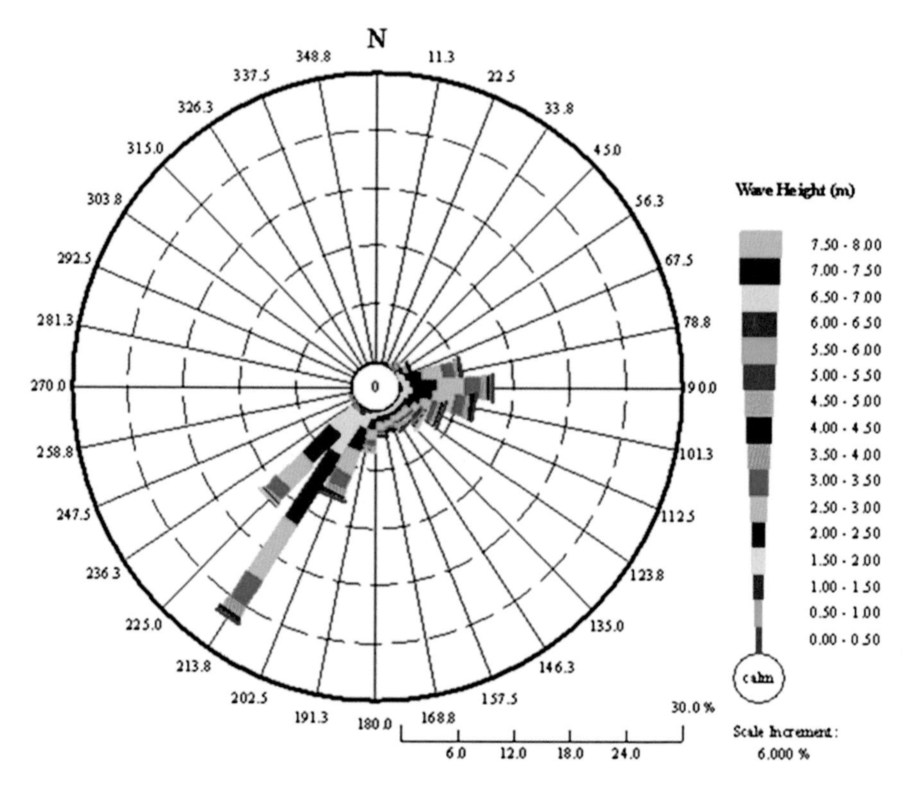


**Fig. 13.6** Wave heights (offshore)

**Fig. 13.7** Wave direction and current from eastern part of the study area (*Source*: QMM)

Usually, dredging activity depends on wave direction approaches. Wave direction approaches to the shore are important for coastal stability because the changes in angle lead to coastal erosion by removing beach sediment and transporting it into the sea. Normally, wave direction approaches should be perpendicular to the shore (Hyndman 2006). In that case, the energy produced by wave forces dissipates into

**Fig. 13.8** Wave direction and current from southern part of the study area (*Source*: QMM)

**Fig. 13.9** Image showing the actual state of Libanona in Galions Bay, 2012

the terrestrial zone and longshore current follows a parallel direction along the shore line. But if wave direction approaches are not perpendicular, a huge amount of energy is dissipated to the shore, leading to coastal destabilization. Hence, coastal zone is unable to resist this energy, and eventually longshore current direction at an angle along the shore line dredges sediment (Hyndman 2006). In Fort Dauphin, coastal erosion may have been caused by the irregularity of wave direction approaches because they were not perpendicular to the shoreline during the fieldwork in 2011 and 2012.

Dredging is always happening in coastal areas but its intensity is weak in the normal environment (Hyndman 2006). Coastal dredging, characterized by the lack of sediment supply or sand in the coastal zone, is probably due to the formation of canyons in the deep sea or atmospheric air disorders. When canyons have formed, nature tries to fill the gaps at the expense of beaches or sensitive areas whose sand can be transported. Both formation of canyons and atmospheric air disorders might have occurred in Fort Dauphin because human-induced dredging during the Ehoala port construction could be one of the causes leading to coastal erosion in this area.

The dredging of the ocean bottom on a superficies of 181,000 m$^2$ (QMM S.A. 2009) might also have led to coastal erosion in Galions Bay between 2004 and 2009. This might also explain the beach diminishing in this area during that period as no major change was noticed at the other beaches.

Furthermore, waves and tides act on cliffs formed of solid rock in the same way on beaches and dunes, but their action is focused at the base and on arches. Some pieces of rock at these points are washed away each time leading finally to instability, or even to rocks being torn off and thrown into the sea at the foot of the cliffs. These rocks are later pulled into the sea, accelerating the erosion. It means that the results of wind-induced erosion and current-induced erosion are the same but the processes are different.

Moreover, during this period (2004–2009) the study area was threatened many times by tropical cyclones. As a result, heavy damage to the coastline was recorded between 2009 and 2011. The damage observed on the beach portions probably resulted from heavy waves with very active currents. In addition, traces of subaerial erosion characterize some places. Coastal erosion and particularly subaerial erosion could therefore be occurring at the same time with heavy rain. Overall, a loss of ten meters from the coastal area was measured between 2009 and 2011. These facts suggest that only cyclones are capable of causing storm surges, winds, and currents at the highest levels that can destroy 10 m of coastal area in 2 years.

Looking at the weather events that have occurred recently in Madagascar, the island has experienced frequent tropical cyclone passage during the past ten years (Rakotondravony 2012) (Fig. 13.10). Generally, tropical cyclone passage is accompanied by heavy rain and violent storms. Tropical cyclones that attacked the country were strongly associated with marine movements in the immediate vicinity of Fort Dauphin. Although the path of a storm may not directly affect the coastal city of Fort Dauphin, this area sustains heavy rains, high waves, and storm surges along the coast (Donque 1974; Météorologie Nationale 1975; Direction de la météorologie 1984). Moreover, these factors have an influence on the local atmospheric circulation and the influence remains even a few years after the cyclone event. It takes approximately 4–6 years before normal conditions return (Nicholson 1997), but due to climate change that disrupts the air circulation, normal conditions may never be recovered (Rasmusson and Wallace 1983). It is also surprising that many tropical cyclones affecting Madagascar, eventually reach Fort Dauphin and dissipate there, or nearby.

During the period from 2004 to 2009 coastal accretion occurred although longshore currents did not significantly affect the beach; the waves were certainly less aggressive due to less precipitation. Nevertheless, residents along the shoreline found that the sea level increased significantly after the tsunami event in 2004. According to villagers, they noticed that the width of the beach had declined by about 80 m since 2004. One factor that can amplify the action of cyclone is a rise in sea level. This takes place extremely slowly and seems minimal, but also causes the removal of shoreline.

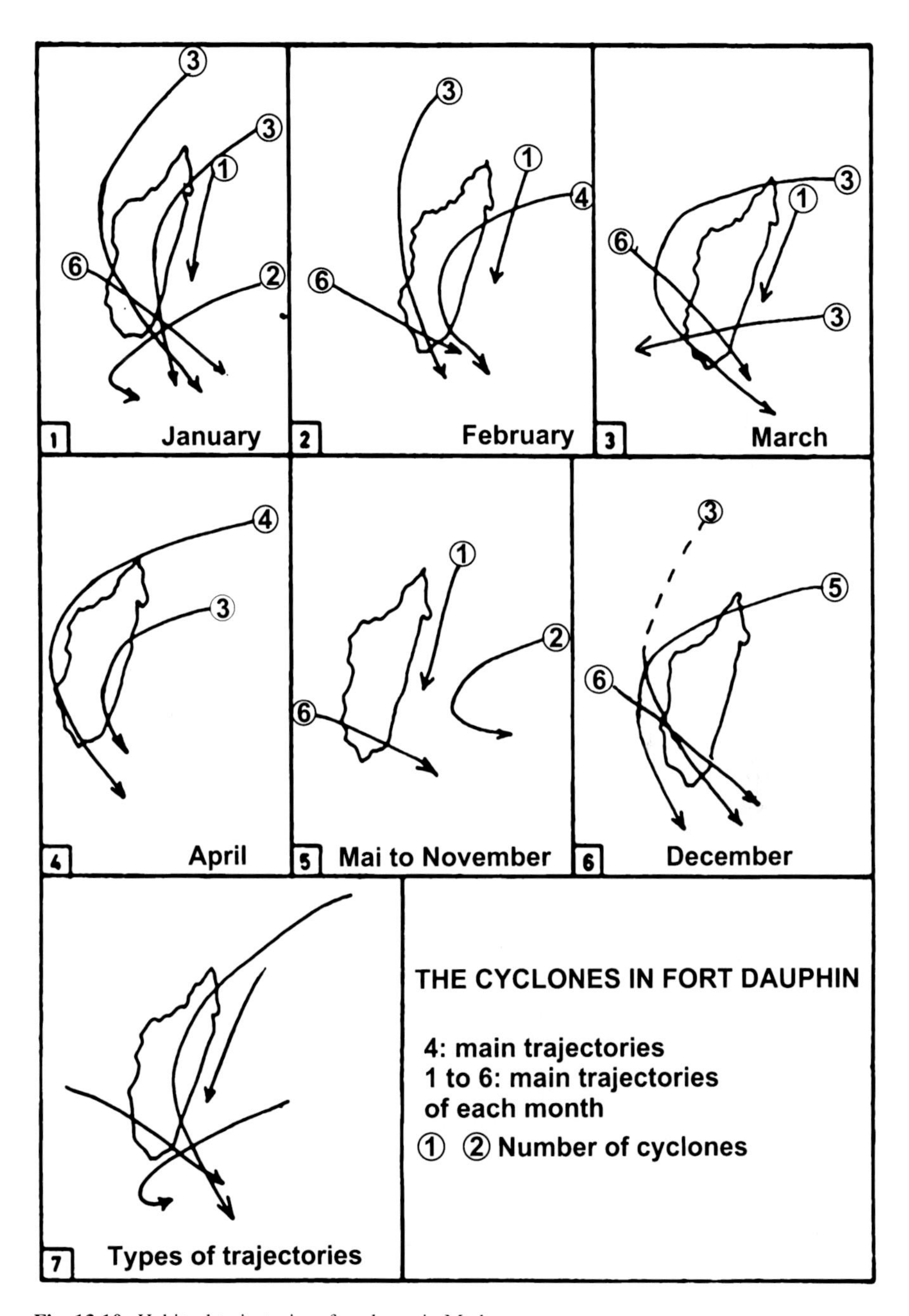


**Fig. 13.10** Habitual trajectories of cyclones in Madagascar

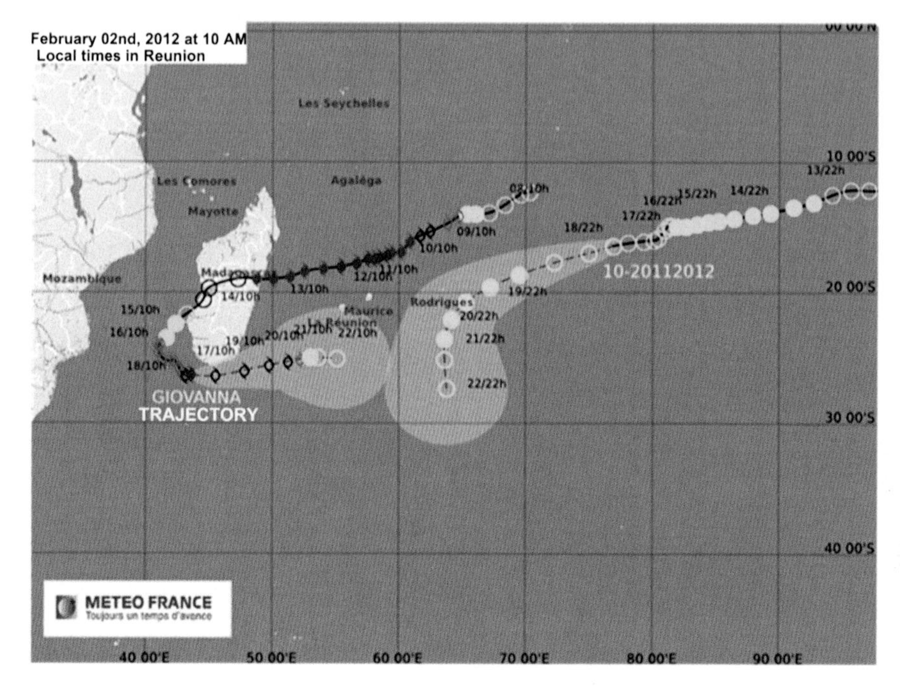


Fig. 13.10 (continued)

## 13.4 Solutions

To limit the damage caused by coastal erosion, residents have devised solutions using gabions or garbage. These initiatives may be collective or individual. A protective wall has been constructed facing the sea on a portion of the coast in the south through a private initiative. Groins have also been installed in the study area by QMM. However, no shoreline protection project has yet been initiated to fight against or mitigate coastal erosion in Fort Dauphin.

As explained above, marine erosion around the city of Fort Dauphin can take different forms (marine, subaerial, or wind erosion) but all are controlled directly or indirectly by the wind. It would be wise to consider several options for countermeasures depending on the type of erosion.

### *13.4.1 Marine Erosion*

Several solutions can be adopted to fight against coastal erosion. However, these solutions must take into account the specifics of protected areas (e.g., tourist sites, areas characterized by dune formations, etc.), as well as the local availability of protective materials. Taking these considerations into account, two or three options were adopted immediately after the damage was sustained to stop or mitigate the effects of coastal erosion in Fort Dauphin. But they were mostly private or individual initiatives.

#### 13.4.1.1 Groins

This technique is based on the construction of stone or concrete walls perpendicular to the shore and extending into the water. They are primarily intended to promote the accumulation of sand on the beach by trapping the sand, or slowing its movement along the shore. They can also be constructed of either wood or steel. The groins can be built separately or in groups along the beach (Fig. 13.11).

These coastal groins will therefore be used to reduce the transport of coastal sand and/or sediment to maintain the current speed away from the shore at high tide. In this case, the wave attack is first concentrated on the end of the groins. Groins operate more efficiently on coasts where the directions of littoral transport are constant. They are one of the most widely used countermeasures against coastal erosion in the world. The best results have been shown by the spikes in Nevis and the Caribbean islands. In Madagascar, they have already proven their effectiveness in protecting the new Ehoala port where accumulations of sand on both sides were observed (Fig. 13.12).

The structures, which soften the action of the waves on the shore, are used on all Fort Dauphin's bays. However, in the case of Libanona, the location of the structures and possible intensity of ocean currents in this small space could further reduce the space available for recreational activity. The beach is the closest recreational area to the city and the most popular beach among the residents.

**Fig. 13.11** Groins installed close to the Ehoala port

**Fig. 13.12** Image showing the groins installed close to the Ehoala port from space

### *13.4.2 Subaerial Erosion*

To mitigate the subaerial erosion, in addition to setting up protective walls at the cliffs, an adequate drainage system needs to be established in Fort Dauphin. Moreover, the current provision for sanitation should be improved, particularly in sensitive areas (near the coast).

### *13.4.3 Wind Erosion*

Wind erosion areas are mainly on the sides of the dunes, and now this erosion also happens on the ridges. This means that the action of wind erosion has reached its peak and may not extend back to the dunes. The affected areas are all unconsolidated dunes. Planting of plant species capable of growing rapidly and transmitting their roots deep into the dunes is proposed. For this solution to be effective, the wind direction on the areas to be protected must be taken into consideration during implementation, as well as the most favorable times to avoid wind washout of the dunes later. At the same time, siltation should be reduced to stabilize the soil. Preliminary

**Fig. 13.13** Casuarinas trees planted near the Ehoala port

steps such as mulch techniques are necessary to prepare the ground before the actual planting of plant species. This entails covering the sand with a uniform layer to stop the action of the wind on the ground and especially to block siltation. First, a fixative (mulch) must be prepared with straw, local herbs, or agricultural residues, and the area subject to erosion must be covered until revegetation occurs. Regular watering of the soil is then undertaken to create cohesion between the grains, making soil much more resistant to erosion.

Once the dunes are stable, planted trees can withstand the actions of the wind over time. Use of Vetiver has already shown great success in watershed protection in Madagascar. This species is capable of sinking its roots up to three meters deep and grows almost everywhere. The same is true for casuarinas, which have already been planted in Fort Dauphin along the road leading to the Ehoala port as a kind of windbreak (Fig. 13.13). To date this species has grown well and met expectations.

## 13.5 Conclusion

Recently this municipality has been threatened by a rise sea level, landslides, and coastal erosion. These natural phenomena have worsened since 2004, resulting in damage to and loss of agricultural land, houses, roads, and recreational beaches. In 2011 and 2012 widespread diminishing of beaches was observed around Fort Dauphin, with the southern areas severely damaged. A combination of satellite imagery analyses and climatic survey enabled us to identify development of coastal erosion in Fort Dauphin. These analyses indicated that average loss of coastal area was about 5 m/year, but coastal erosion activity became stronger during the period between 2004 and 2011.

In this area, coastal erosion was characterized by a combination of coastal, subaerial, and wind erosion. This combination occurs especially during rainy seasons and in particular during cyclones. Indeed, the wave height is much greater during these periods, and it is the same for the ocean currents and rainfall. In general, coastal erosion is at its peak when the effects of the three types of erosion (coastal, subaerial, and wind erosion) are combined.

Causes of coastal erosion in Fort Dauphin might result from a combination of factors, both natural and human-induced, which have different patterns in time and space. They are probably due to the cyclone passage that has generated storms and tides over recent years in Madagascar. Simulations of waves and their height indicate that south-west and easterly wave directions are responsible for the damage. Human activity such as dredging and river damming in the south and north of the study area might have contributed to the intensity of coastal erosion during 2004–2011. The movements of these parameters depend on wind direction and dynamism.

The PIC project made it possible to understand the key processes of coastal dynamics and how coasts function both in spatial and temporal time scales, allowing solutions to fight against coastal erosion to be proposed. Meanwhile, some solutions have already been implemented mostly as individual or collective initiatives. These initiatives have demonstrated the efficacy of constructing groins and planting species such as vetiver or casuarinas trees, with accretion resulting in some places.

The combination of offshore and onshore techniques such as groins and vegetation in coastal areas will improve slope stability, consolidate sediment, and diminish the amount of wave energy moving onshore, thereby protecting the shoreline from erosion.

## References

Auckland Regional Council (2000) Coastal erosion management manual in coastal hazard strategy. Technical Publication No. 130. ISSN: 1175 205X

Brebbia CA, Benassai G, Rodriguez G (2009) Coastal processes. In: First international conference on physical coastal processes, management and engineering. WIT Southampton, Boston. ISBN: 978-1-84564-200-6

Cambers G (1998) Planning for coastline change. 2b shoreline management in Nevis. A Position Paper

DGENV European Commission (2004) Development of guidance. Document on Strategic Environmental Assessment (SEA) and coastal erosion. Final Report

Direction de la météorologie (1980) La saison cyclonique de 1979–1980 à Madagascar. Direction de la météorologie, Mad Rev De Géo No. 38. Antananarivo Madagascar, p 72 (in French)

Direction de la météorologie (1984) La saison cyclonique de 1983–1984. Service de La Réunion, Mad Rev De Géo No. 43. Antananarivo Madagascar, p 75 (in French)

Donque G (1974) Le climat d'une façade au vent de l'Alizée: la côte Est de Madagascar. Mad Rev De Géo 24, Tananarive, p 55 (in French)

Hyndman (2006) Natural hazards and disasters. Hyndman and Hyndman. ISBN: 0-534-99760-0. http://earthscience.brookscole.com/hyndman

Ir Zamani Bin Mindu (1988) Coastal erosion: problems and solutions. In: Proceedings of the 11th annual seminar of the Malaysian society of marine sciences

Kim YC (2010) Handbook of coastal and ocean engineering. World Scientific Publishing, Singapore

Météorologie Nationale (1975) La saison cyclonique de 1974–1975. Service de La Réunion, ronéoté, Mad Rev De Géo No. 27, p 28 (in French)

Neuvy G (1981) Aménagement régional à Madagascar. Morondava, un cas de l'érosion Marine. Mad Rev De Géo 38, pp 67–88 (in French)

Nicholson SE (1997) An analysis of the ENSO signal in the tropical Atlantic and western Indian Oceans. Int J Climatol 17:345–375

Niesing H (2005) EUROSION: coastal erosion measures, knowledge and results acquired through 60 studies. National Institute for Coastal and Marine Management, Public Works and Water Management, Ministry of Transport, The Hague, pp 421–431

NSW Government (2010) Coastal planning guideline: adapting to sea level rise. State of New South Wales through the Department of Planning. www.planning.nsw.gov.au. ISBN: 978-1-74263-035-9. Pub no. DO P 10_022

Prasetya GS, Black KP (2003) Sanur and Kuta Beaches in Bali – case studies for replacing traditional coastal protection with offshore reef. In: Proceedings of the artificial surfing reef, Raglan

QIT Madagascar Minerals S.A. (QMM S.A.) (2009) PGES – Port d'Ehoala et Transport Routier – Phase Opération – Version finale (in French)

Rakotondravony H (2012) Rapport sur l'Etat de l'Environnement à Madagascar. Chapitre 10: Catastrophes Naturelles (in French)

Rasmusson EM, Wallace JM (1983) Meteorological aspect of the El Nino/Southern Oscillation. Sciences 222(4629):1115–1202

Ratsivalaka Randriamanga S (1985) Recherches sur le climat de Taolagnaro (ex-Fort-Dauphin) (Extrême sud de Madagascar). Mad Rev De Géo No. 46, pp 46–67 (in French)

Rio Tinto-QMM (2008) PGES Bilan social et environnemental: addenda – sols in Port d'Ehoala et Transport Routier – Phase Opération – (in French)

Schiereck GJ (2004) Introduction to bed, bank and shore protection. Spon Press; Taylor and Francis, London and New York, Published in the USA and Canada

Shore Protection Manual (1984) Department of the Army Waterways Experiment station, Corps of Engineers. Coastal engineering researcher center. 4th edn. vol 2. U.S. Government Printing Office, Washington, DC

Slovinsky PA (2011) Maine Coastal Property Owner's Guide to Erosion, Flooding, and Other Hazards (MSG-TR-11-01). Maine Sea Grant College Program, Orono

Wang Y (2010) Remote sensing of coastal environments: an overview. In: Wang Y (ed) Remote sensing of coastal environments. CRC, Boca Raton, pp 1–21

Williams AT, Micallef A (2009) Beach management: principles & practice. First published by Earthscan in the UK and USA in 2009

# Part III
# Degradation of Environment and Mitigation

# Chapter 14
# Risk Management of Chemical Pollution: Principles from the Japanese Experience

**Shigeki Masunaga**

**Abstract** This chapter discusses some experiences of environmental pollution management in Japan. Cases include air pollution, water pollution, and toxic chemical regulation. From those experiences, it is concluded that in-process management of pollutants has been the major cause of pollutant reduction. End-of-pipe technologies, such as flue gas and wastewater treatment, are costly and not as efficient as cleaner production processes. In addition, governmental policies in environmental management, such as setting regulatory values and guiding industries, sometimes lead industries to take inefficient countermeasures. Thus, regulating agencies should take care to ensure that their policy is leading society the right way and have the flexibility to adapt their policy as necessary.

**Keywords** End-of-pipe treatment • In-process management • Japanese experiences of environmental pollution • Toxic chemical regulation

## 14.1 Introduction

Japan's economic development was very rapid during the 1960–1970s, when the country was recovering from the economy's devastation as a result of World War II. As industrial development progressed, environmental pollution became evident and caused severe health damage to the people living around some industrial areas. These included four infamous cases of pollution-related health damage that resulted in major legal suits, namely, Minamata Disease around Minamata Bay in Kumamoto and Agano River in Niigata, Itai-Itai Disease on Jinzu River in Toyama, and

S. Masunaga (✉)
Faculty of Environment and Information Sciences, Yokohama National University, 79-7 Tokiwadai, Hodogaya-ku, Yokohama 240-8501, Japan
e-mail: masunaga@ynu.ac.jp

N. Kaneko et al. (eds.), *Sustainable Living with Environmental Risks*,
DOI 10.1007/978-4-431-54804-1_14, 

Yokkaichi Asthma around Yokkaichi petrochemical complex in Mie. In these cases, the national and local governments were afraid of inhibiting economic development and were at first reluctant to take measures against the pollution and to compensate for the health damage. The victims therefore had to appeal to the courts to assert their rights. Indeed, the people affected by the pollution and the non-governmental groups that supported them played important roles in obtaining court decisions in favor of the victims, forcing the government and parliament to take countermeasures.

It has often been said that Japan successfully reduced pollution without hindering its economic growth. In my opinion, this statement is partially true, but not totally; there have been both merits and demerits in the way Japan acted. In this article, I would like to describe how the Japanese government and industries coped with the pollution problems and discuss what can be learned from Japanese experiences both good and bad.

## 14.2 How Pollution Loads Were Reduced

### *14.2.1 Air Pollution in Yokkaichi Petrochemical Complex*

A large scale petrochemical industrial complex commenced operations in Yokkaichi City around 1960. It contributed about one fourth of total Japanese petrochemical production value in the early 1960s. In 1961, however, people in the surrounding area started to complain of asthma and the situation reached its worst level around 1963–1964. In those days, the sulfur content of heavy oil was about 3 % and the emission of sulfur oxides as $SO_2$ was about 130,000–140,000 tons/year (Study Group on Global Environmental Economics 1991). In the Isodu area south of the complex, about 3 % of the one-hour-average $SO_2$ concentration exceeded 0.5 ppm, which was 5 times higher than the current environmental standard, a one-hour-average concentration of 0.1 ppm. In 1967 the residents of the Isodu area, where the air pollution was most severe, appealed to the courts to stop the air pollution and to obtain compensation for their health damage. The courts acknowledged joint tort of the companies in the petrochemical complex and judged in favor of the plaintiffs. As a consequence, the following countermeasures were taken:

1. Introduction of environmental quality standards and stricter emissions regulation by the national government. The first regulation of individual companies' emission loads was introduced in Yokkaichi.
2. Shift to heavy oil with lower sulfur content in the petrochemical industry.
3. Construction of heavy oil desulfurization plants.
4. Installation of flue gas desulfurization equipment in the petrochemical industry.

As a result of these measures, the targeted environmental standard for $SO_2$ was attained by 1976. Annual average $SO_2$ concentrations at monitoring stations in Yokkaichi are shown in Fig. 14.1. Estimated emissions of $SO_2$ in the Yokkaichi area

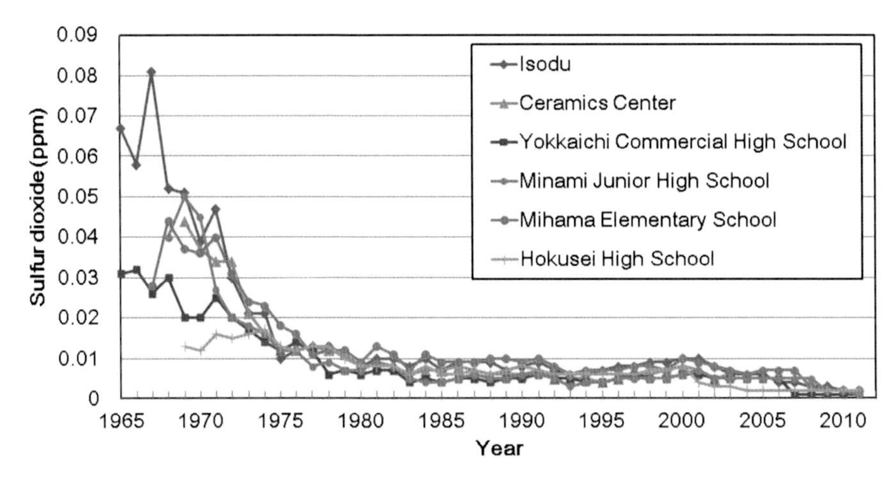


**Fig. 14.1** Annual sulfur dioxide concentrations at monitoring stations in Yokkaichi City (based on data from Yokkaichi City (2000, 2012))

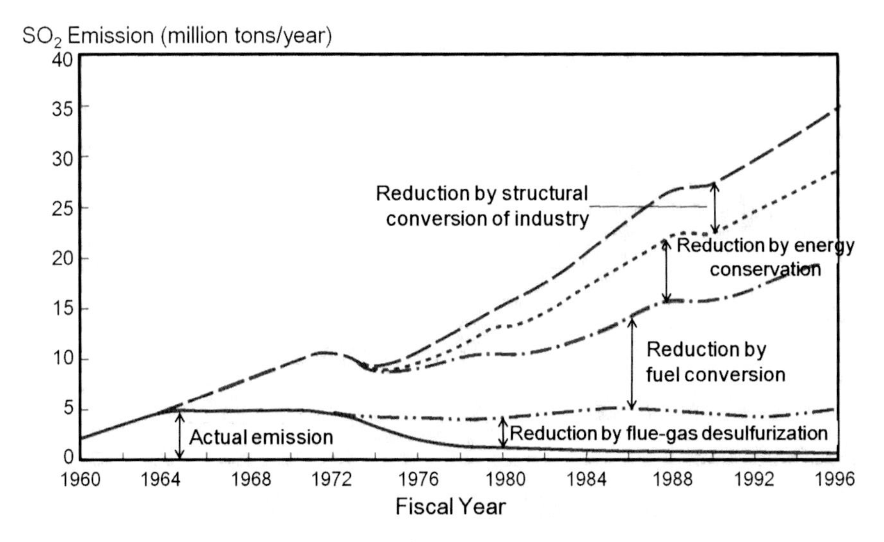


**Fig. 14.2** Contribution of different factors that reduced sulfur dioxide emissions in Japan (Committee on Japan's Experience in the Battle Against Air Pollution 1997)

were drastically reduced from 140,000 tons/year in 1964 to 100,000 in 1971 and to 20,000 in 1975. After the late 1970s, gradual reduction continued. In contrast, fuel consumption in the area was relatively constant during the same period (about 4 million kL/year) (Mie Prefecture 1990). This situation could be explained by the analysis for the whole of Japan undertaken by the Committee on Japan's Experience in the Battle Against Air Pollution, as shown in Fig. 14.2 (Committee on Japan's Experience in the Battle Against Air Pollution 1997). This figure shows that fuel

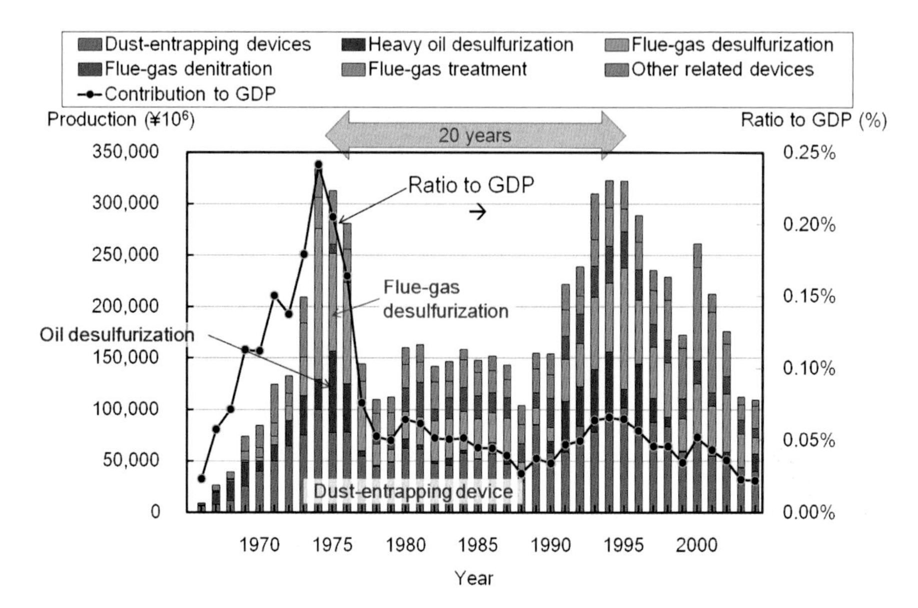


**Fig. 14.3** Actual production values of air pollution prevention devices (*bar graph*) and their contributions to GNP (*line graph*) in Japan (FY1966–2004). Based on Statistics Japan (Statistics Bureau 2005)

conversion (shift to oil of lower sulfur content and desulfurization of fuel) was the major cause of the reduction, followed by energy conservation. Flue gas desulfurization made only a small contribution. This experience taught us that reduction at source is important and effective compared to end-of-pipe flue gas treatment.

The petrochemical sector had to invest a considerable amount of money to achieve the reduction. Figure 14.3 shows the production of air pollution prevention devices in Japan. The first installation peak appeared around 1974 and this corresponds to the rapid improvement of air quality discussed earlier in this section for the case of Yokkaichi. The largest amount of money was used for installation of flue-gas desulfurization facilities, followed by dust-entrapping devices and oil desulfurization. It should be noted that investment in oil desulfurization facilities was minimal, but it was much more cost effective in terms of reducing emissions. However, to reduce emissions further, flue-gas desulfurization had to be installed. Overall, the most effective technology may vary according to the region and the nation's economic situation; which technology to use therefore needs to be reconsidered in each individual case. As a percentage of gross national product (GNP), investment in devices to prevent air pollution was nearly 0.25 % in 1974. This high number shows that Japan made quite a concentrated investment to counteract air pollution during the 1970s. The second peak of investment in air pollution prevention devices occurred in the mid-1990s, twenty years later. This was when the facilities built in the 1970s were updated. The second peak was almost as high as the first peak in terms of nominal value; however, the contribution to GNP was much smaller—dropping to as little as 0.07 %—due to the growth of GNP during the 20 years.

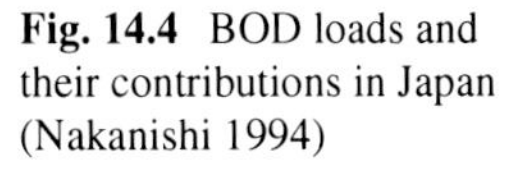

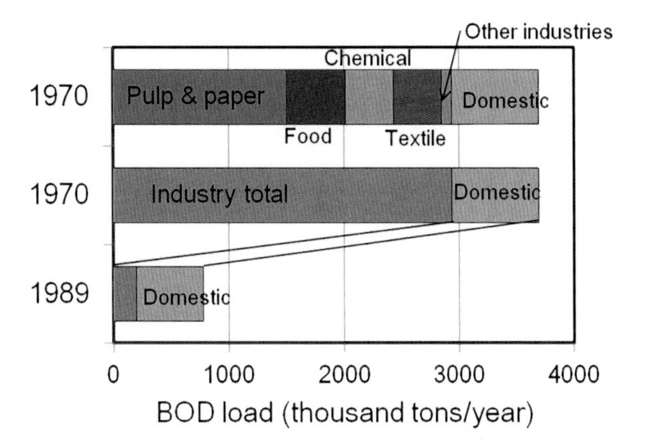


**Fig. 14.4** BOD loads and their contributions in Japan (Nakanishi 1994)

### *14.2.2 Preventing Water Pollution in the Pulp and Paper Industry*

The total organic pollutant load in terms of biochemical oxygen demand (BOD) from domestic sewage was reduced from 750 thousand tons/year in 1970 to 580 thousand tons/year in 1989 (Nakanishi 1994). This reduction was to only three-fourths of the original amount. In comparison, the total BOD discharged into public water bodies as wastewater from all Japanese industries was significantly reduced to about one fifteenth of its original value, from 3 million tons/year (in 1970) to 200 thousand tons/year (in the late 1980s) (Fig. 14.4). As the organic pollutant load from the pulp and paper industry was quite large and contributed about half of the total BOD load from all industries, here we will discuss reduction of the wastewater organic pollutant load from the pulp and paper industry in particular.

The organic pollutant load from the pulp and paper industry was about 2.2 million tons/year in terms of chemical oxygen demand (COD) in 1970, and it was reduced to only 200 thousand tons/year in 1989. Assuming no reduction measures had been taken, however, it was estimated that it would have been 4.5 million tons/year in 1989 based on the amount of pulp and paper production (Nakanishi 1994). The COD reduction was attained mainly by a product shift (58 %) and by black liquor recovery (26 %) (Fig. 14.5). The pulp and paper companies shifted their product from pulp with high unit COD emission load to that of low unit emission load such as Kraft pulp. In addition, waste paper was used as a raw material in place of virgin pulp, which also reduced the emission load.

Black liquor is the spent cooking liquor that is formed when pulpwood is digested into paper pulp by removing lignin, hemicelluloses, and other extractives from wood. This liquor contains a very high concentration of organic pollutants that are hard to remove by biological wastewater treatment processes. Thus, instead of discharging black liquor to treatment plants or directly into water bodies, companies started to concentrate the liquor and burn it to recover heat energy to generate steam

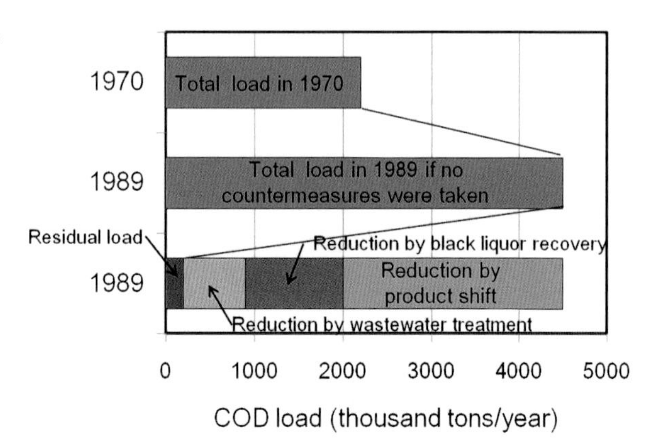


**Fig. 14.5** Reduction of COD load in pulp and paper industry (Nakanishi 1994)

and cooking chemicals (sodium hydroxides and sodium sulfide). This process change not only reduced the COD emission but was economically beneficial in terms of energy and chemical use.

The third reduction measure was wastewater treatment, which contributed about 16 % of the total reduction. It should be emphasized, however, that the reduction of waste load by improving the production process was much more effective and economical than end-of-pipe wastewater treatment. In-process reduction also often saved raw materials and energy.

In contrast to industrial wastewater, domestic wastewater load cannot be reduced at the source but is instead reduced by municipal wastewater treatment plants or by small community-based sewage treatment systems. These sewage treatments are usually more expensive and take a longer time to implement than industrial wastewater treatment measures as long sewer pipes have to be installed.

### *14.2.3 Environmental Standards for Toxic Chemical Substances*

As stated in the introduction section, Japan experienced some cases of environmental pollution that caused damage to human health. Because of those experiences, the Japanese government started to regulate a limited number of toxic pollutants that caused problems, such as mercury, cadmium, lead, and PCBs, by setting strict effluent standards (Table 14.1). As a result of these measures, rates of compliance with environmental water quality standards were significantly improved during the 1970s. The number of regulated pollutants in water did not change, however, for about 20 years until the 1990s. In 1993, fifteen new substances including organochlorines and pesticides were added to the items subject to the environmental quality standards. Three more items were added in 1999 and one item was added in 2009. The reason that itemized regulation developed so slowly was that setting an

**Table 14.1** Items subject to water quality standards during the 1970s in Japan

| Items | Standard (original value) | Year first introduced (revised) |
|---|---|---|
| Cadmium | 0.01 mg/L | 1970 (2011: 0.01 → 0.003 mg/L) |
| Cyanide | ND | 1970 |
| Lead | 0.1 mg/L | 1970 (1993: 0.1 → 0.01 mg/L) |
| Hexavalent chromium | 0.05 mg/L | 1970 |
| Arsenic | 0.05 mg/L | 1970 (1993: 0.05 → 0.01 mg/L) |
| Total mercury | 0.0005 mg/L | 1970 |
| Alkyl mercury | ND | 1970 |
| Organophosphate | ND | 1970 (1993: deleted from the list) |
| PCB | ND | 1975 |

*ND* Should not be detected
The number of items regulated for protection of human health is currently 27

environmental standard for each item meant extensive and costly monitoring as well as control of emissions in a range of industrial sectors across the nation. It was therefore difficult to get support from all the sectors to add new items. Faced with the difficulty of increasing the number of items subject to standards, in 1993 the Japanese government designated a group of chemicals that should be monitored (currently 26 items) and another group of chemicals that should be under observation (about 300 chemicals). Chemicals in these new groups are not yet regulated by the law but they are being monitored or information is being collected about them. Based on the information collected, new pollutants will be selected for inclusion among the items subject to environmental standards. This type of soft regulation gives warning to industry that, although some of the chemicals they use are not currently regulated, those chemicals should be used with care. Another mechanism that provides companies with an incentive to reduce their pollutant emissions is the pollutant release and transfer register (PRTR) system, whereby companies are requested to report the amounts of listed pollutants they discharge into the environment or transfer to wastewater and solid waste treatment facilities.

### *14.2.4 Case of Administrative Guidance Relating to Toxic Chemicals*

The chlor-alkali industry produces sodium hydroxide (caustic soda) and chlorine, as well as hydrogen, by electrolysis of sodium chloride solution. It is one of the key industries in the chemical sector. There are a few different methods, but the mercury cell process employing inorganic mercury was the main one used in Japan after World War II.

In 1956 the first Minamata disease patient was officially identified in the city of Minamata and in 1959 organic mercury was found to be the cause of the disease by Kumamoto University researchers. In 1965, a second case of Minamata disease was identified in the downstream area of Agano River. In both cases, discharge of organic mercury from the acetaldehyde production process, rather than the chlor-alkali production process, was officially confirmed as the cause of the disease by the national government in 1968.

However, mercury was also emitted from the chlor-alkali production process, which was linked to a third case of Minamata disease reported in the Ariake Sea area in 1973. This case in particular caused great fear among the Japanese public, although the link with chlor-alkali production was later disproved. As a result of the public outcry, in June 1973 the Japanese government decided to minimize mercury loss from the mercury cell process by introducing a closed system. In addition, it decided in November 1973 that one third of Japan's mercury cell operations would be converted to the diaphragm cell process by September 1975, with the remainder to be converted by March 1978. Although at that time it had not been established that inorganic mercury could be transformed into organic mercury in the environment, a committee appointed by the government had submitted a report saying that the mercury problem would not be solved without totally phasing out the mercury cell process. Consequently the Japanese government made the drastic decision of ordering the chlor-alkali industry to change their production process.

Following the emergence of the chlor-alkali industry in the 1920s, the production capacity of the mercury cell and diaphragm cell processes became comparable during the 1930s. This situation continued until the mid-1950s, when demand for high quality sodium hydroxide increased. The mercury cell process could produce better quality sodium hydroxide, but mercury was expensive; the diaphragm cell process, on the other hand, produced an inferior quality product, but without the need for mercury. When the Japanese chlor-alkali industry was converting its mercury cell operations to diaphragm cell operations, it became evident that the sodium hydroxide produced by the diaphragm cell process contained about 1 % sodium chloride and other impurities such as sodium chlorate. This made it unusable for the production of chemical fiber, cellophane, inorganic chemicals and so on, which account for about 25 % of total sodium hydroxide demand (Kameyama 2008). Its chlorine gas was not suitable as a raw material for products such as vinyl chloride, which account for 41 % of total demand (Kameyama 2008). In addition, the prices of diaphragm cell products were expected to be 40 % more expensive than those of mercury cell products. To promote conversion from one process to the other, therefore, the Japanese government had to align the prices of corresponding products from the two processes by introducing a system for paying the price difference. The resulting lack of incentive to innovate was not favorable considering that international competition and innovative technology were desired.

Subsequently, following the appearance of an innovative ion-exchange membrane cell process developed by Asahi Glass Co., completion of the conversion plan's second stage was postponed to await confirmation of the applicability of the innovative technology. The new process, which could produce better quality products with high energy efficiency, was adopted instead, and accounted for 60 % of Japan's production capacity in 1987 and 100 % in 2005. This process of transition within the chlor-alkali industry is shown in Fig. 14.6.

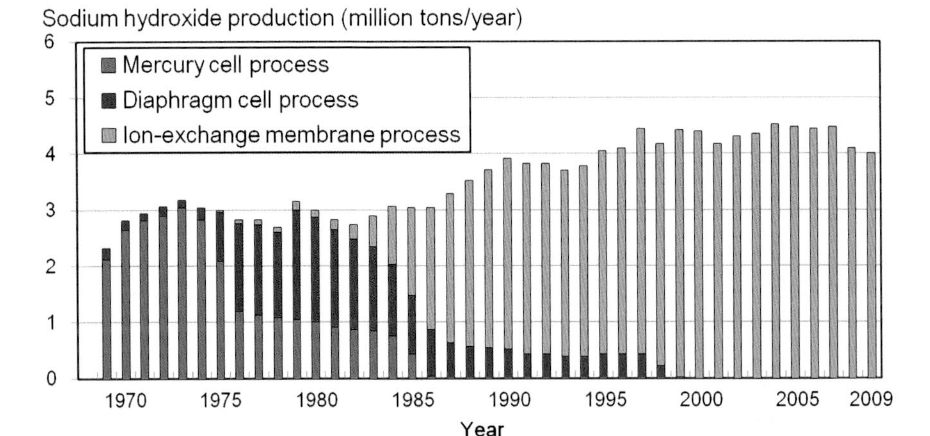


**Fig. 14.6** Transitions in chlor-alkali processes in Japan (Ministry of the Environment and Japan 2011)

The conversion between processes in Japan's chlor-alkali industry was a peculiar case, but it was instructive. The Japanese government had been compelled to take quite a precautionary policy after Minamata disease had such a large impact on society. However, its administrative guidance conflicted with economic principles and could not be adopted by the industry without a price adjustment system. The chlor-alkali industry had to waste a significant amount of money to change their process first from mercury cell to diaphragm cell, and then to the membrane cell process. The Japan Soda Industry Association estimated that the chlor-alkali industry invested about 10 years-worth of benefit for the process conversion in 1978 alone (Nakanishi 1995). In fact, the closed-system mercury process reduced mercury loss from 114 to 2.3 g per ton of sodium hydroxide production and, with hindsight, was good enough to bridge the time-gap until the membrane process emerged (Kameyama 2008). Delaying the transition would have been much more cost efficient. On the other hand, it can be said that without the government order, the industry might not have worked hard enough to develop an innovative technology in such a short period.

It is likely that the Japanese chlor-alkali industry could bear the high cost of process change in the 1970s and 1980s because they faced less pressure from foreign countries' imports than is the case now that world trade is more open. It is becoming increasingly difficult for industries to bear the increase in costs for better environmental management if a similar degree of management is not required in other countries. At any rate, management policies should be readjusted to fit the current situation so that both environmental protection and economic development can progress.

## 14.3 Conclusion: Lessons from the Japanese Experience

This chapter introduced some examples of environmental pollution and associated countermeasures in Japan. Successful pollution reductions are often achieved by in-process technological innovation and efforts, not by wastewater or flue gas treatment. In-process pollution reductions are referred to as "cleaner production," and are often much more economical than end-of-pipe treatment. In addition, wise water use, such as recycling and counter flow washing, contributes to raising the concentration of pollutants in the water and making treatment more efficient. Perhaps the key lesson to be drawn from the Japanese experience, however, is that governmental regulation and guidance should be introduced carefully and adaptively.

**Acknowledgment** The author referred to the books and reports written by Dr. Junko Nakanishi (Nakanishi 1994, 1995), Dr. Tetsuya Kameyama (Kameyama 2008), Dr. Takamitsu Sawa (Committee on Japan's Experience in the Battle Against Air Pollution 1997), and others to prepare this manuscript.



## References

Committee on Japan's Experience in the Battle Against Air Pollution (1997) Japan's experience in the battle against air pollution: working towards sustainable development. The Pollution-Related Heath Damage Compensation and Prevention Association, Tokyo (Currently, Environmental Restoration and Conservation Agency, Kawasaki City, Japan)

Kameyama T (2008) Chemical technology and environmental issues – Minamata disease and caustic soda production. Sci Net 32:6–9

Mie Prefecture (1990) Environmental white paper: the environment of Mie Prefecture

Ministry of the Environment, Japan (2011) Lessons from Minamata disease and mercury management in Japan. http://www.env.go.jp/chemi/tmms/pr-m/mat01/en_full.pdf

Nakanishi J (1994) Environmental strategy for water. Iwanami Shoten Publishers, Tokyo

Nakanishi J (1995) Environmental risk theory: policy proposal from technology arguments. Iwanami Shoten Publishers, Tokyo

Statistics Bureau (2005) Statistics Japan: Production of environmental equipment. http://www.stat.go.jp/data/chouki/30.htm

Study Group on Global Environmental Economics (1991) Experiences of environmental hazards in Japan - Uneconomy of non-consideration to the environment. Godo Shuppan Co. Ltd, Tokyo

Yokkaichi City (2000) Environmental protection in Yokkaichi City. http://www5.city.yokkaichi.mie.jp/menu73747.html

Yokkaichi City (2012) Environmental protection in Yokkaichi City. http://www5.city.yokkaichi.mie.jp/menu80127.html

# Chapter 15
# Research on the Correlation Between Chlorophyll-a and Organic Matter BOD, COD, Phosphorus, and Total Nitrogen in Stagnant Lake Basins

**Song Toan Pham Phu**

**Abstract** We are frequently required to assess, mitigate, and monitor certain environmental risks involved in our daily activities at both the local and global environment levels. Corporations, in particular, can cause environmental pollution or other risks as a result of either accidents or natural events. In order to limit, and hopefully prevent, these environmental impacts, environmental risk management places a strong emphasis on targeting the problems that could arise, and implements a system of metrics that help with prevention and management. Eutrophication is one example of a widespread environmental pollution phenomenon that is currently worsening in lake basins. Proliferation of phytoplankton is due primarily to high concentrations of nitrogen (N) and phosphorus (P) discharged from industrial and urban waste. However, assessing, managing, and forecasting the pollution of basins requires considerable time and support tools to analyze a range of water quality indicators over time and geographical area. This study investigated the relationship between organic matter (chemical oxygen demand, biochemical oxygen demand, total N, and P) and biomass of phytoplankton (indicated by Chlorophyll-a) in order to shorten the period required for analysis, predict eutrophication of lakes ahead of time, and promptly prevent the spread of contaminants. Based on this relationship, we can use Chlorophyll-a as a biological indicator in monitoring and assessing the levels of organic pollution. In addition, we can identify ways to reduce pollution and transportation of pollutants in stagnant lake basins, and contribute to reducing the damage due to environmental risks.

**Keywords** Chlorophyll-a • Environmental risk management • Eutrophication • Organic pollutants • Phytoplankton

---

S.T.P. Phu (✉)
The University of Da Nang, Da Nang, Vietnam
e-mail: ppstoan@gmail.com

N. Kaneko et al. (eds.), *Sustainable Living with Environmental Risks*,
DOI 10.1007/978-4-431-54804-1_15, 

## 15.1 Introduction

Water pollution is a problem of concern all over the world and it involves difficult issues relating to water management and environmental protection. It not only affects a large water surface and destroys aquatic communities, but also impacts on human health and quality of life. In developing countries nowadays, the consequences of industrial development and urban population growth involve issues of water pollution, leading especially to pollution of foodstuffs. This raises the urgent need for a monitoring program with specific early warning targets to enable measures to be taken to prevent pollution in a timely manner.

Eutrophication is a natural process that has been occurring for thousands of years. Its speed has increased rapidly in recent decades due to human activities. Eutrophication involves an increase in nutrients (especially nitrogen and phosphorus) in water, causing excessive growth of the lower species of plants such as algae. It creates major changes in aquatic ecosystems, causing deterioration of water quality. The biochemical basis of the eutrophication process is photochemical reaction, which increases phytoplankton biomass (Fig. 15.1).

There are many indicators to assess water quality including physical, chemical, and biological indicators. Phytoplankton in particular are considered to be an evaluation parameter for organic pollution. Chlorophyll, meanwhile, has long been known as a major pigment with the role of absorbing solar radiation energy as part of the optical response of plant biomass. Therefore, assessment of phytoplankton biomass through chlorophyll analysis requires a combination of chemical and biological methods. From chlorophyll concentration we can establish the relationship between chlorophyll and other pollution parameters to be able to quickly assess the status of eutrophication and organic pollution in a lake basin. Thus we can evaluate water quality through chlorophyll indicators.

**Fig. 15.1** The eutrophication process (*Source*: www.renault.com)

**Table 15.1** OECD classification of nutrients in lakes

| Condition of lake | Chl-a (mg/m$^3$) | Chl-a$_{max}$ (mg/m$^3$) |
|---|---|---|
| Lack of nutrients | <1 | <2.5 |
| Poor nutrients | <2.5 | <8.0 |
| Average nutrients | 2.5–8.0 | 8.0–25.0 |
| Eutrophication | 8.0–25.0 | 25.0–75.0 |
| Super eutrophication | >25 | >75 |

*Source*: OECD-Organization for Economic Co-operation and Development

The role of phytoplankton in nutrient-rich freshwater is large. They provide oxygen for aerobic organisms, disintegrate organic compounds, and remove minerals and other nutrients from the environment. But when excessive growth of phytoplankton occurs, algal blooms reduce the decomposition of algae in the sediment and lead to the destruction of water ecosystems.

There are many methods of assessing phytoplankton biomass, including:

- Calculating individual density combined with measuring average individual volume of all species;
- Measuring the weight of all individuals in a unit volume;
- Measuring carbon concentration derived from organic matter;
- Measuring the pigment concentration;
- Measuring the rate of exchange of oxygen and carbon dioxide;
- Measuring adenosine triphosphate (ATP).

Of the above, measuring pigment concentration is the most viable method. The photosynthetic reaction of phytoplankton generates biomass, which can therefore be assessed just by determining the pigment concentration. This differs from the other parameters above that are integrated in the metabolism of phytoplankton. Therefore analysis of Chlorophyll-a (Chl-a) is a good means of estimating the nutritional status of a lake. This method is simpler, faster, and more economical than estimating phytoplankton biomass by microscopy. Thus Chl-a concentration can determine the level of lake eutrophication (Table 15.1).

## 15.2 Subjects of Study

### *15.2.1 Chlorophyll*

Chlorophyll is the common name of a plant pigment capable of absorbing solar energy for the process of plant photosynthesis. Within plant cells, chlorophyll is organized distinctively, distributed in cytoplasm called chloroplasts. The molecular structure of chlorophyll comprises two parts: the porphyrin ring and the phytol (an

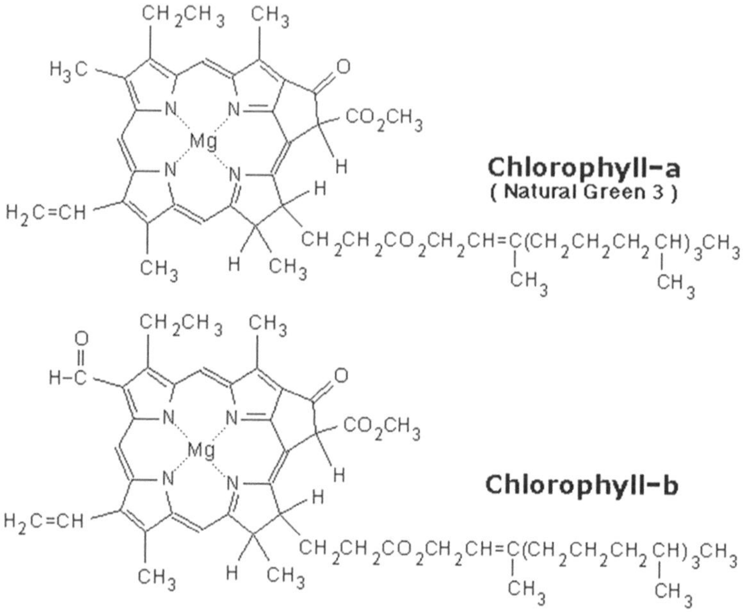

Fig. 15.2 Structure of Chlorophyll-a and Chlorophyll-b (*Source*: www.ysi.com)

acyclic hydrocarbon). The porphyrin ring is a strong chelating ligand created when four N atoms strongly associated with a metal atom (Mg) are coordinated in a planar arrangement (Fig. 15.2).

In nature, the two most common forms of chlorophyll are Chlorophyll-a ($C_{55}H_{72}O_5N_4Mg$) and Chlorophyll-b ($C_{55}H_{70}O_6N_4Mg$) with a ratio of Chlorophyll-a to Chlorophyll-b of 3:1.

There are several methods to determine chlorophyll:

- UV–vis photometry
- Fluorescence measuring
- High Pressure Liquid Chromatography (HPLC)
- Use of remote sensing images

### 15.2.2 BOD: Biochemical Oxygen Demand

BOD is the amount of oxygen demand for microorganisms to oxidate and stabilize dissolved organic or inorganic substances in water under certain conditions. Bacteria use dissolved oxygen (DO) for decomposition of organic matter and as a result DO will be reduced.

The oxidation of organic matter in water can occur in two phases:

- The first phase is mainly oxidation of hydrocarbon (HC). This phase occurs for 20 days at 20 °C

$$HC + O_2 + \text{Microorganism} \rightarrow CO_2 + H_2O$$

- The second phase is oxidation of nitrogen-containing compounds. This phase starts from the tenth day (or can start as early as the fifth day).

$$2NH_3 + O_2 + \text{Microorganism} \rightarrow 2NO_{2^-} + 2H^+ + 2H_2O$$

$$2NO_{2^-} + O_2 + \text{Microorganism} \rightarrow 2NO_{3^-}$$

BOD is the single most important parameter to assess water pollution. The higher the BOD concentration, the more biodegradable organic matter exists. A typical concentration might be "$BOD_5$" which is oxygen demand to oxidate biodegradable organic matter at 20 °C for 5 days.

## 15.2.3 COD: Chemical Oxygen Demand

COD is the amount of oxygen demand to oxidate organic matter fully in water. It is a parameter for both organic matter easily susceptible to biodegradation and organic matter that resists biodegradation. Thus, COD is determined by the photometric method with strong oxidizing reagents in an acid environment and involvement of $Ag_2SO_4$ as a catalyst.

$$C_nH_aO_bN_c + dCr_2O_7^{2-} + (8d + c)H^+$$
$$(Ag_2SO_4) \rightarrow nCO_2 + (a + 8d - 3c)/2H_2O + cNH^{4+} + 2dCr^{3+}$$

This method allows COD to be determined from 10–500 mg/L in 2 h.

## 15.2.4 Phosphorus

Phosphorus exists in water in many forms: ortho phosphate ($PO_4^{3-}$), polyphosphate ($P_2O_7^{4-}$, $P_3O_{10}^{5-}$) and organic phosphorus. In addition, bio-inorganic and organic phosphorus is found in sediment and sludge. Phosphorus is one of the nutritional components necessary for the life of organisms, especially aquatic life.

Ortho phosphate is determined by the photometric method with a vanadate molybdate reagent.

$$PO_4^{3-} + VO_{3^-} + 11MoO_4^{2-} + 25H^+ \rightarrow H^3PVMo_{11}O_{40} + 11H_2O$$

### 15.2.5 Nitrogen

Nitrogen exists in water in several forms:

- Ammonium ($NH_4^+$): The metabolic product of nitrogen-containing compounds in wastewater.

$$NH_3 + H^+ \rightarrow NH_4^+$$

- Nitrate ($NO_3^-$): In a low pH environment, $NH_3$ is metabolized to $NH_4^+$ and oxidized to $NO_3^-$ by oxygen.

$$NH_4^+ + 2O_2 \rightarrow H_2O + 2H^+$$

- Nitrite ($NO_2^-$) The intermediate product of the nitrogen cycle, nitrite is toxic to fish and other aquatic life.

Determination of nitrogen is based on the principle of oxidizing samples to produce the ammonium form of nitrogen.

$$NH_4^+ + OH^- \rightarrow NH_3 + H_2O$$

$$HBO_2 + NH_3 = NH_4^+ + BO_{2^-}$$

Ammonium is determined using titrate $BO_2$- with HCl 0.01N:

$$BO_{2^-} + H^+ + = HBO_2$$

## 15.3 Research Methodology

### 15.3.1 Determination of Chlorophyll

#### 15.3.1.1 Chlorophyll Refining

The chlorophyll refining process is as shown in Fig. 15.3. Chl-a was extracted from bamboo leaf in 3 h using the Soxhlet system with acetone as the solvent. The products obtained were evaporated to remove the solvent and Chl-a was tested for purity using plate chromatography with an ultraviolet light wavelength of 254 nm.

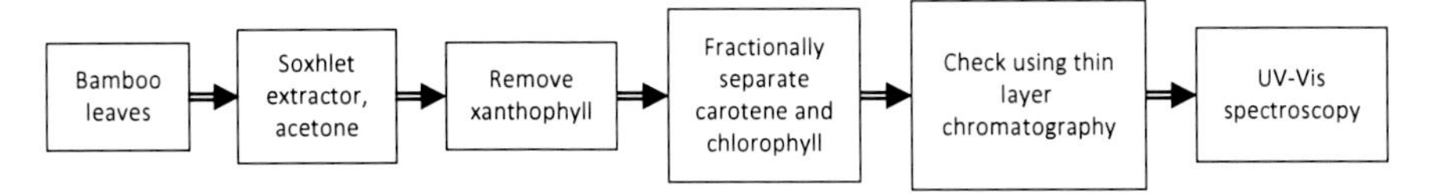


**Fig. 15.3** The chlorophyll refining process

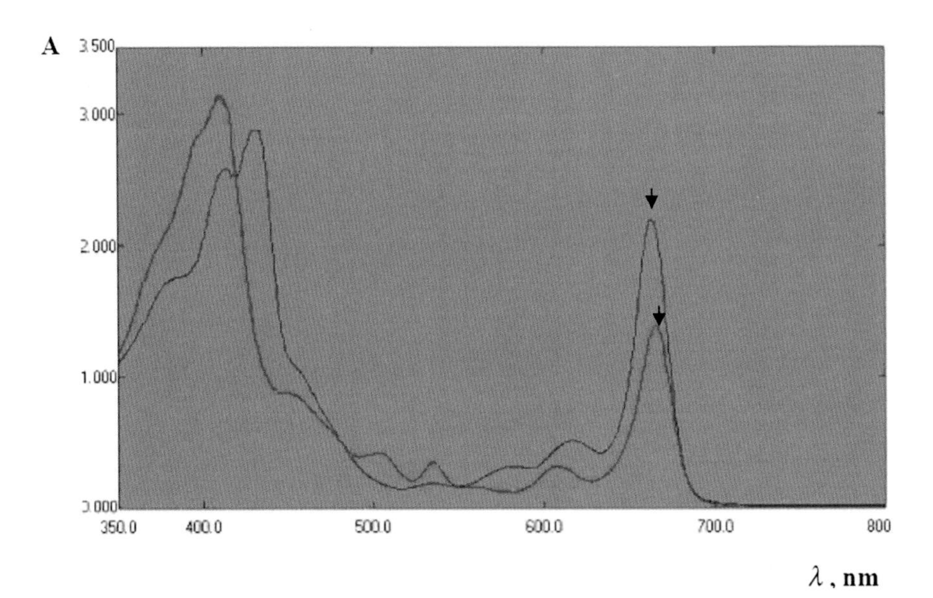


**Fig. 15.4** UV–vis absorption spectra of Chl-a (*top*) and pheophytin-a (*bottom*)

#### 15.3.1.2 Determining the Maximum Absorption of Chl-a

Refined chlorophyll was dissolved in acetone. The resulting liquid was measured using UV–vis spectroscopy. Spectrometric results showed that the maximum absorption in the visible region corresponded to wavelengths of 664 and 665 nm for pheophytin-a and Chl-a (Fig. 15.4). Thus, a 664 nm wavelength was chosen to measure the absorbance in the determination process.

#### 15.3.1.3 Building a Chl-a Calibration Curve

Dissolve purified chlorophyll in a mixture of 90 % acetone and 10 % saturated $MgCO_3$ and prepare norms of standard volume at a concentration of 100,000 g/L.

Dilute the working Chl-a liquid at a concentration of 5,000 μg/L.

Build the ranges of standard fluids from the working Chl-a and measure absorbance at a wavelength of maximum absorbance of Chl-a: $\lambda = 664$ nm Fig. 15.5 depicts the calibration curve of Chl-a in chart form.

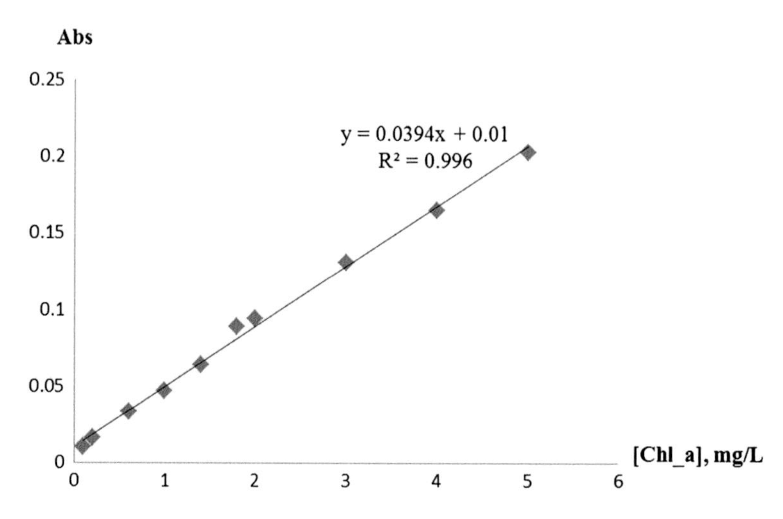


**Fig. 15.5** Calibration curve of chlorophyll-a

## *15.3.2 Determination of BOD$_5$*

| Prepare dilutions: | – 1 L of oxygen saturated solution<br>– 1 mL of buffer liquid<br>– 1 mL of $Mg^{2+}$<br>– 10 mL of $Ca^{2+}$<br>– 10 mL of $Fe^{2+}$ |
|---|---|

- Put $V_d$ (mL) sample into BOD bottle
- Control pH at around 7
- Add enough V (mL) dilution liquid to fill BOD bottle
- Measure $DO_i$ using DO meter
- Cap BOD bottle and put in BOD cabinet at 20 °C for 5 days
- Measure $DO_f$ by DO meter after 5 days

BOD$_5$ can be calculated using the following formula:

$$BOD_5 = \frac{(DO_i - DO_f).V_d}{V_s}$$

where $DO_i$ is the amount of dissolved oxygen in the diluted sample; $DO_f$ is the dissolved oxygen content of the sample after 5 days of incubation at 20 °C; $V_d$ is the volume of the sample after diluting; $V_s$ is the volume of the sample before diluting.

### *15.3.3 Determination of COD*

#### 15.3.3.1 Building a COD Calibration curve

- Prepare standard fluids; mix standard fluids at concentrations of:

10 20 40 60 80 100 120 140 (mg/L)

- Put exact volumes of standard fluids into the ampoules: (Table 15.2)
- Shake ampoules well, dry them, and put them into the heating machine at 150 °C for 120 min
- Centrifuge at a speed of 3,000 cycles/15 min
- Do the same for the fresh sample using 2.5 mL of distilled water instead of the standard fluid

Figure 15.6 depicts the calibration curve of COD in chart form.

**Table 15.2** Volumes of standard fluids in ampoules

| Number | Fluids | V (mL) |
|---|---|---|
| 1 | Sample | 2.5 |
| 2 | $K_2Cr_2O_7$ | 1.5 |
| 3 | $Ag_2SO_4$ | 3.5 |

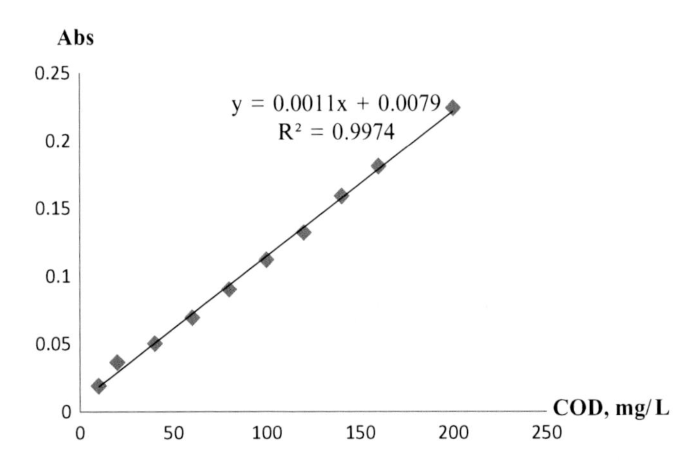


**Fig. 15.6** Calibration curve of COD

### 15.3.4 Determination of Phosphate

Building a phosphate calibration curve:

- Prepare standard fluids; mix standard fluids with $PO_3^{4-}$ concentrations of:

  0 0.4 2.0 4.0 6.0 8.0 10.0 (mg/L)

- Put 5.0 mL vanadate-molybdate into each bottle of fluid; norms to be 50 mL by distilled water; shake thoroughly
- Wait for 10 min at room temperature; measure the absorbance at 400 nm wavelengths

  A calibration curve can be built in the same way as for COD.

## 15.4 Results

### 15.4.1 Correlation Between Chl-a and COD

The research subjects were Ham Nghi Lake and March 29 Park Lake, both located in the city of Danang (Figs. 15.7 and 15.8). The study period was from March to May 2013 (the dry season).

**Fig. 15.7** Ham Nghi Lake (*Source*: http://rongbay.com/Da-Nang/)

Fig. 15.8 March 29 Park Lake (*Source*: http://wikimapia.org/4681749/)

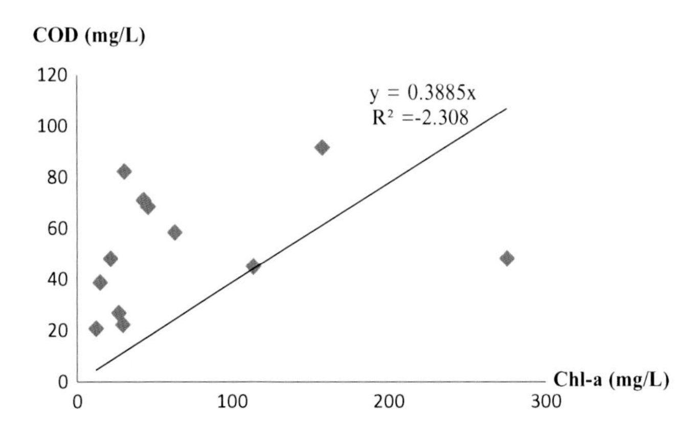


Fig. 15.9 Correlation between Chl-a and COD in Ham Nghi Lake

By comparing the results for COD and Chl-a, we can assess the correlation between them (Figs. 15.9 and 15.10). No correlation was identified between the concentrations of Chl-a and COD. This was because COD (chemical oxygen demand) is a pollution parameter for all components including organic compounds that resist biodegradation. The growth of phytoplankton is related only to organic pollution.

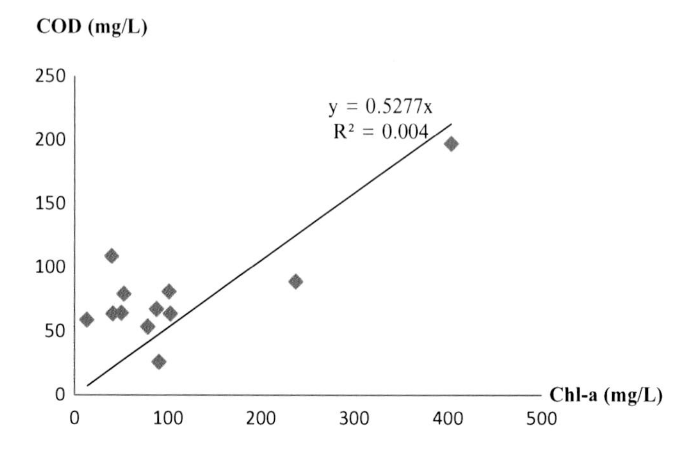


**Fig. 15.10** Correlation between Chl-a and COD in March 29 Park Lake

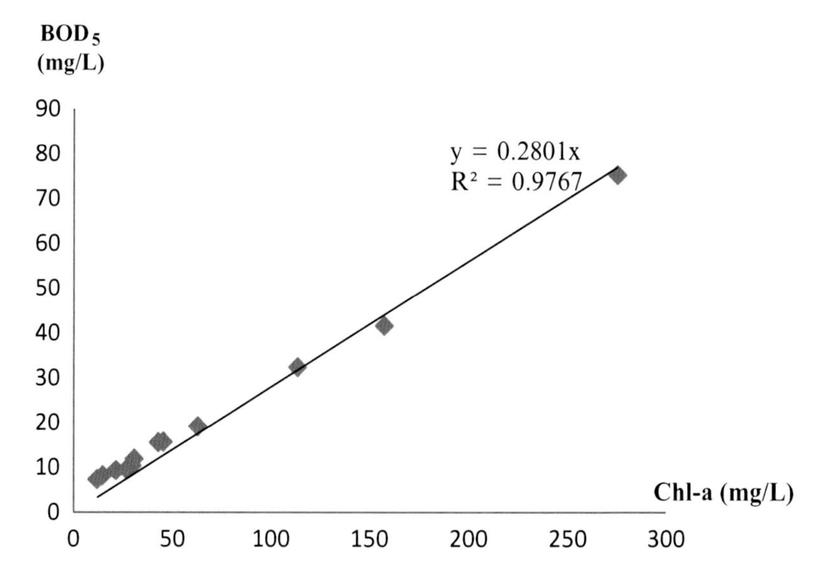


**Fig. 15.11** Correlation between Chl-a and BOD in Ham Nghi Lake

## *15.4.2 Correlation Between Chl-a and BOD*

BOD was analyzed at the same time as COD. The BOD analysis results are shown in Figs. 15.11 and 15.12.

We can see that BOD and Chl-a have a clear correlation with a reliable linear gain coefficient of $R^2 > 0.95$.

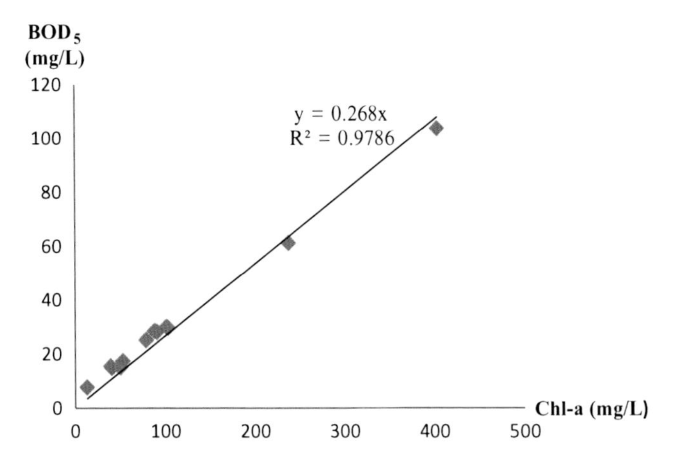


**Fig. 15.12** Correlation between Chl-a and BOD in March 29 Park Lake

In both research subjects, there is a high linear correlation between the concentrations of Chl-a and BOD.

### *15.4.3 Correlation Between Chl-a and Phosphate and Total Nitrogen*

Biological cycles of nitrogen and phosphorus occur in water as they do on land. However, due to the light attenuation and water stratification, biological processes that absorb and replicate nutrients occur differently by depth. The phenomenon of water stratification and seasonal changes varies in different climate zones. Stratification causes temperature changes and changes in the salinity of the water. When a water body is stratified, the exchange between surface and depths is very slight. The process of temperature change creates a significant barrier to the spread and transport of substances between two layers of water, so nutrients in the top layer can be depleted and limit algal productivity.

The study period was from March to May 2013, and samples were taken from approximately 30 cm below the water surface. The correlation charts are shown in Figs. 15.13 and 15.14.

The result of observation in the two lakes during three months in 2013 indicated no clear correlations between the parameters of phosphate or nitrogen content and of Chl-a. This result could have been affected by weather factors such as rain, the depth of sampling locations, and sunlight, which make algae either float on the water or sink.

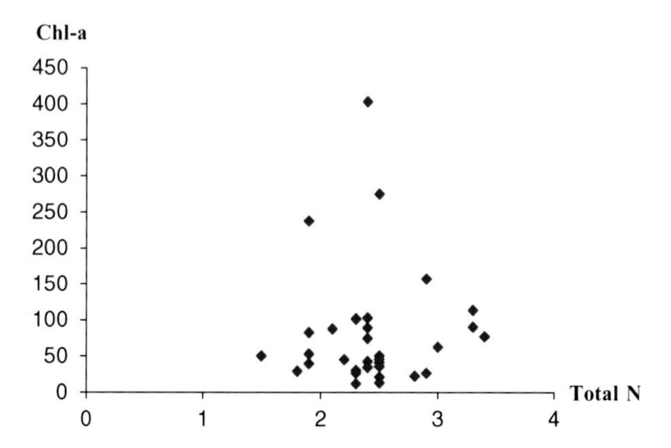


**Fig. 15.13** Correlation between Chl-a and phosphate

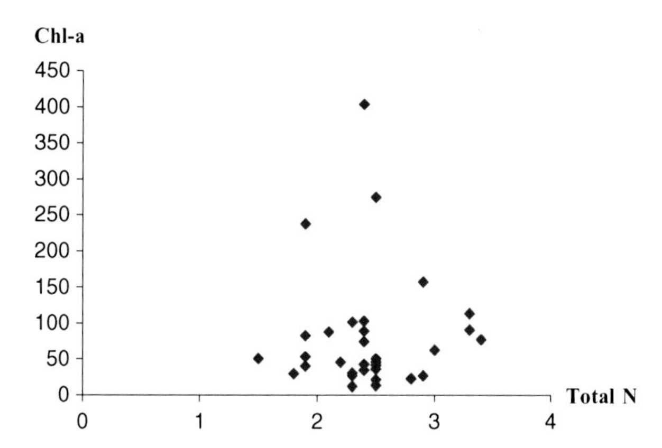


**Fig. 15.14** Correlation between Chl-a and total nitrogen

## 15.5 Conclusion

Chl-a is an important parameter in water quality assessment, as the growth of phytoplankton depends significantly on the concentration of organic matter in the water. However, the research shows:

- No correlation between Chl-a and COD as is usually the case in standing surface water
- A clear linear correlation between Chl-a and BOD in both research subjects:

$$\text{Ham Nghi Lake : Chl} \quad a = 0.2848\,x + 3.7344 \text{ with } R^2 = 0.955$$

$$\text{March 29 Park Lake : Chl} \quad a = 0.2435\,x + 5.097 \text{ with } R^2 = 0.998$$

– Phosphate and total nitrogen are two important factors directly affecting the development of algae and they are also causes of eutrophication. However, no correlation was found between Chl-a and phosphate or total nitrogen during the research period (March–May 2013).

Management of water and water pollution is an important task within environmental risk management. Use of Chl-a as an ecological indicator to assess water quality is required in order to optimize the tools of water quality management, as well as to quickly detect and reduce the damage from environmental risk. It should be combined with modeling tools to forecast, monitor, and provide scenarios of environmental risk and global climate change.

## 15.6 Research Orientation

Further research is needed to:

– Investigate the relationship between Chl-a and nutrient parameters such as total nitrogen and phosphate over the long term and in more research subjects under comparable weather conditions (during the dry and wet seasons)
– Identify the impact of climate factors, weather, and seasonal variation on the nutrient content of the water body
– Study wider basins such as rivers with many contributing factors to find out the general relationships between Chl-a and the above parameters to apply to any watershed
– Develop research in unstable basins to forecast dispersion of pollution and the eutrophication process
– Examine the use of Chl-a as an environmental indicator to assess water quality

# Chapter 16
# Managing Construction Development Risks to the Environment

**Nik Norulaini Nik Ab Rahman and Norizan Esa**

**Abstract** The control of environmental risks arising from construction has become a major issue for the public. Environmental risk is defined as any risk or potential risk to the environment (of whatever degree or duration) and includes all types of impacts. Much of the research conducted has portrayed construction as a major contributor to environmental disruption and pollution. Pollution risks due to construction are typically classified as air pollution, waste pollution, noise pollution, and water pollution. Controlling the risks demands the ability to manage these types of pollution or eliminate their generation. The endorsement of environmental risk management and the mission of sustainable development have resulted in pressure demanding the adoption of proper methods to improve environmental performance in the construction industry. This chapter therefore places its focus on the risks arising from the construction industry, and how to manage these risks so as to strike a balance between development and environmental concerns. Real examples of construction activities are briefly presented to enable readers to envision the risks and the actual efforts undertaken to curb them. This chapter also presents the perception of the public at large regarding the risks caused by construction.

**Keywords** Construction-related pollution • Environmental risks of construction • Sustainable construction

N.N.N.A. Rahman (✉)
School of Distance Education, Universiti Sains Malaysia,
Penang 11800, Malaysia
e-mail: norulain@usm.my

N. Esa
School of Educational Studies, Universiti Sains Malaysia,
Penang 11800, Malaysia
e-mail: norizanesa@usm.my

N. Kaneko et al. (eds.), *Sustainable Living with Environmental Risks*,
DOI 10.1007/978-4-431-54804-1_16, 

## 16.1 Introduction

Construction is not an inherently environmentally friendly industry and much research has linked it to environmental disruption and pollution. Construction is one of the major contributors to environmental impacts carrying pollution risks, typically classified as air pollution, waste pollution, noise pollution, and water pollution. Activities in construction are complex, highly dispersed, and resource-demanding. The industry contributes to the loss of important natural assets and imposes severe impacts and stress on the environment. Construction activities and practices that fail to control risks to the environment can cause damage to rivers, lakes, sensitive ecosystems, and aquatic life, including causing fish kills. They can also disrupt wildlife habitats and result in contamination of land and groundwater (Sanvicens and Baldwin 1996). The risks to the environment are particularly high when work is undertaken on highlands, steep slopes, and near coastal areas, rivers, and lakes.

In Malaysia, the construction industry has yet to achieve full sustainability, and the environment is frequently being irretrievably altered, while risks are poorly handled. A marked example is in Penang by the tendency to re-designate optimum locations or sensitive zones such as fish landing grounds for upscale developments. In light of the potential consequences, site selection needs to be undertaken with greater thoroughness and deliberation, including efforts should be made to save energy via construction of green buildings. Green buildings are specially designed to cause less risk to human health and the environment through efficient use of energy, water, and other resources, with a concurrent reduction in waste, pollution and environmental impacts.

According to Tam et al. (2004), sustainable construction requires the local construction industry to be viable, capable, and alert to the imperatives of sustainable construction in all its activities. Input from those who can make changes in the industry, such as developers, contractors, architects, planners, and others must be provided in a cohesive and willing manner (Gangolells et al. 2011; Christini et al. 2004).

## 16.2 Environmental Risks of Construction

### *16.2.1 Risk of Land Degradation*

Large projects usually entail extensive land disturbance involving the removal of vegetation and reshaping of topography. Such activities make the soil vulnerable to erosion. Soil removed by erosion may become airborne and create dust problem or be carried by water into natural waterways, thereby polluting them. Due to the soil erosion of the exposed and loose earth, there will be a deterioration of water quality in the surrounding water bodies due to siltation. This can result in mud floods and flash floods in immediate or downstream areas during heavy downpours. Landslides and slope failure can occur on unstable slopes or when the soil is saturated with water during heavy rainfalls.

### 16.2.2 Risk to Flora and Fauna

The biological environment includes various species of animal and plant life, and their habitats. Unfortunately, however, loss of flora and fauna is likely in any development. Planning is essential to ensure minimal losses during the implementation stages and steps must also be taken later to ensure that the losses are "replenished." This is crucial especially when development is in the vicinity of a protected or animal sanctuary, a forested area, or a catchment area. Ecological losses should be minimized and suitable protection put in place for the conservation of flora and fauna.

### 16.2.3 Risk of Water Pollution

Water quality is important for economic, ecological, aesthetic, and recreational purposes. Changes in water quality may affect its aesthetic value or even prevent some uses of the water. During construction, the potential for soil erosion and risk to water quality is greatest when removal of vegetation for initial clearing and grading activities exposes soil and makes it susceptible to erosion. The impacts and risks are greatest during the rainy season where extensive land clearing can increase sediment load into the rivers from erosion of the exposed soil.

### 16.2.4 Risk of Air Pollution

Activities of major concern for air quality are the burning of waste, the emission of fumes and smoke, and the release of chemical impurities such as heavy metals, acid and other toxic bases. Air quality impacts from construction include increased dust particulates in the atmosphere caused by grading, filling, removals, and other construction activities. Air quality may also be impacted by emissions from construction equipment and vehicles.

### 16.2.5 Risk from Noise and Vibration

Noise and vibration can be generated by various activities and equipment used in construction projects. Noise and vibration levels due to construction activities in the project area vary depending on the types of equipment used, the location of the equipment, and the operating mode. Adverse impacts resulting from construction noise and vibration are generally limited to areas adjacent to the project, and are temporary in nature.

## 16.3 How to Determine Risks

The first thing organizations involved in construction need to do is determine the environmental risks that they have to manage. Environmental risks are those risks that are associated directly and indirectly with the construction activities (Põder 2006). In the construction sector, the environmental risks can be one or more of the types described in 16.2 above. These can happen during the course of the construction, during the operation, or at the end of operation, as well as in abnormal or emergency situations such as during heavy rain or landslides. These environmental risks should therefore be taken into consideration at every stage of the project implementation, from conception to completion (Hickie and Wade 1997). A project can give rise to a number of environmental risks related to the activities of the various organizations involved in construction. Some risks will be directly within the organizations' control (e.g., direct risks such as air emissions and water discharges) and some will be of a nature that can be only indirectly influenced (e.g., the activities of raw material suppliers) (Olander and Landin 2005). As both types can lead to substantial environmental impacts, they should be assessed for significance. Such risks may be local, regional, or global, short or long term, with varying levels of significance. Having evaluated environmental risks for the significance it is then possible to prioritize actions that address issues relating to the construction operations. Table 16.1 shows the sequence for identifying environmental risks associated with construction.

## 16.4 How to Manage Construction Risks

The construction industry must minimize and manage the generation of pollution that may affect elements of the environment. To ensure that all reasonable and practicable measures have been taken, attention must be given to (a) the nature of the pollution or potential pollution and sensitivity of the receiving environment; (b) the financial implications of various measures that might have to be taken; and (c) the current state of technical knowledge and the likelihood of a successful application of the various measures to be taken (Lee and Fong 2002). For example, general soil erosion and the generation of sediment during construction activities cannot be entirely prevented but should be minimized (Zobel et al. 2002; Zobel and Burman 2004). Sound project planning can reduce the potential for erosion but control measures will always be necessary to reduce the impact of erosion both on-site and off-site. The control measures may consist of a combination of construction strategies, structural and vegetative measures, and soil stabilization techniques. For maximum effect, it is important that all of the control measures to be implemented are integrated into the site development plan (Tam et al. 2004). The complexity and extent of control measures required will depend largely on the magnitude and duration of the construction activity (Olander and Landin 2005). Overall there is a high

**Table 16.1** Sequence involved in risk identification process

| STEP 1. Identify construction type | STEP 2. Identify environment-related activities that can spawn risks | STEP 3. Determine environm-ental risks associated with each activity | STEP 4. Evaluate significance of the environmental risks identified | STEP 5. List all significant activities into a significant risks register | STEP 6. Prioritize the significant risks register. The most significant should be managed first |
|---|---|---|---|---|---|
| • Design of township, highway, or multistory apartment<br>• Land clearing and earth work<br>• Construction of hospital, school, or road | Examples of environment-related activities:<br>• Consumption of natural resources<br>• Removal of top soil<br>• Removal of vegetation<br>• Discharge of waste water, waste oil, building waste, etc.<br>• Consumption of electricity<br>• Emission of dust and other particulates | Examples of environmental risks:<br>• Depletion of natural resources<br>• Loss of soil fertility<br>• Loss of flora and fauna<br>• Loss of habitat<br>• Soil erosion<br>• Visual impact<br>• Water pollution – waste, siltation<br>• Air pollution – dust and particulates<br>• Global warming<br>• Noise pollution<br>• Flash floods<br>Link between activities and risks is similar to "cause and effect" | • Using established methodology and evaluation criteria appropriate to the construction sector<br>• The methodology can be qualitative, quantitative, and semi-quantitative<br>• The evaluation team should have a common understanding of the methodology and application of the criteria<br>• Application of the methodology and evaluation criteria should be consistent throughout the process | • The evaluation process should result in a significant risks register<br>• The register leads the organization to the following:<br>a. The activities, products, or services that need to be managed in order to reduce the environmenal risks<br>b. The activities that are covered under legal and other requirements<br>c. The types of work that require specially trained personnel in order to prevent environmental risks | • It is critical to prioritize within the register to ease the management process<br>• Priority should be based on significance in terms of the environment and legal compliance<br>• Identify measures that can be implemented to prioritize and assuage the risks<br>• The final register will guide the organization in setting its environmental policy and objectives |

level of awareness and commitment with regard to protecting the environment among contractors and others involved in the construction industry. Some have already taken measures to comply with all regulations regarding water pollution, noise control, and dust emissions. In some cases, additional initiatives have also been adopted, such as energy conservation and recycling of materials (Rodríguez et al. 2007; Tam 2008).

## 16.5 Case Studies

### *16.5.1 Case Study 1 Project: Construction of High-Rise Apartment Buildings at Bukit Gambier, Penang*

During this project, quite extensive site clearing took place before actual construction commenced. Environmental risks were incurred during the mobilization, site clearing, and earthworks that required clearing of the ground ready for development. Most of the 8.1 acres of land were located on slopes with contours ranging from 20 to 55 m in elevation and gradients of 5° to >26°. Once the project site had been cleared of vegetation, the disturbed soil surface lost its cohesion. The top soil became very loose as a consequence of tree uprooting, bulldozing, and surface soil disturbances. The increased susceptibility of the soil particles to sheet erosion or displacement by wind or rain was a major concern, since this would increase the volume of silt content in surface runoff during heavy rain. As a countermeasure, surface runoffs and storm water were channeled to an earth drainage system and temporary siltation pond with a perimeter earth drain built to prevent overflow of surface runoffs into adjacent areas. In addition, earthworks were scheduled to be carried out mainly during the dry season and the exposed surface was covered with plastic sheets to avoid rain-wash and siltation/sedimentation. Despite these measures, however, the impacts arising from the project site clearing included the accumulation of overburden, exposed soil leading to extensive silting as a result of surface runoff, and the concomitant degradation of water quality, as well as noise and air pollution from on-site vehicles (Lee and Fong 2002). During drier seasons, exposed soil can also lead to a significant increase in dust suspended in the atmosphere.

### *16.5.2 Case Study 2 Project: Construction of High-Rise Housing at Bayan Lepas, Penang*

The project comprised a total area of 14.1 acres (5.7 ha) at Bayan Lepas with terrain ranging from 10 to 80 m in elevation. The development concept for the proposed project consisted of a medium density residential development of less than 22 units per acre. Since runoff from construction sites can contribute significant sediment loads to receiving water, effective erosion and sediment control at the construction site were part of the measures to manage storm water. In order to minimize risks to the neighborhood from the construction, planning of the development followed an environmentally friendly approach by minimizing cut and fill and maximizing slope protection measures. In addition, sediment control practices were implemented to trap detached soil particles moved by rain, surface run-off, or wind. By implementing Erosion and Sediment Control Practices (ESCP) on site throughout the construction period, the erosion rates were minimized.

### 16.5.3 Case Study 3 Project: Development of Marina and Jetty Infrastructure at Church Street Pier, Pengkalan Weld, Penang

The development site was located in the midst of a regional commercial and tourist hub with existing transportation, communication, and energy infrastructure well in place. There were two types of construction—land-based and sea-based. Site clearing and earthworks within the project area were very limited because the site was primarily an open area covered partly by a concrete platform with some grassy patches and tarred roads. The main activity, was the demolition of the existing pier followed by construction of a new pier, a marina building, and ancillary buildings and structures. The demolition waste included asbestos, zinc planks, wood beams, wiring, glass, and pipes. Waste from the demolition was carefully segregated and disposed of according to type, since major risks are involved with hazardous materials from demolition and associated safety and health issues. Other hazardous waste, such as used oil, was kept in labeled containers to be sent to the designated treatment facility. The sea-based construction including piling disturbed the seabed and released pollutants, leading to an increased water pollution level locally.

### 16.5.4 Case Study 4 Project: Demolition of Part of an Existing Building Structure and Renovation of a Hotel in Batu Ferringhi, Penang

The project plan followed Penang state's principle strategy of developing, protecting, and optimizing land usage in a sustainable way. The project was viewed as an effort to optimize land use as it involved modification of and additions to a commercial development (a hotel) within an existing area and without further expansion. A risk assessment ascribed risk to those activities closely associated with solid waste from the demolition, water pollution of the nearby rivers from the clearing and earthworks, and construction (Tam 2008). This type of development normally imposes less risk than usual; however the local authority had given instructions for an environmental impact assessment to be conducted nonetheless.

The construction site was small and vehicle emissions and dust formation were confined within the construction area by natural vegetative barriers. Vehicles were encouraged to move slowly to reduce air emissions and dust formation. Dry, exposed soil was also sprayed using a water bowser to curb dust formation.

### *16.5.5 Case Study 5 Survey*

A total of 175 adult members of the public were given a questionnaire about the risks posed by the construction industry. This survey revealed that the majority were very much aware of the various risks faced by people, plants, and animals, as well as the risks to the environment as a result of construction activities. For example, most of them agreed that construction activity can cause soil erosion, floods, loss of natural resources, reduction in soil quality, and reduction of rain absorption area. The majority also recognized that loss of habitat, death, and destruction are some of the effects on plants and animals. Similarly, most of them agreed that animals may move away from an area where construction takes place. In addition, a large number of respondents also acknowledged that two risks faced by humans are health problems due to dust released during construction and feeling discomfort due to the noise from construction. However, only around one third of the sample considered that murky water caused by soil erosion from construction sites could have a negative effect on fish in rivers.

In general, most of the people surveyed held similar perceptions about construction risks irrespective of their academic or social background. One notable exception, however, was educational level: Among the sample who responded to this survey, those who had completed university education obtained a mean score of 63.09 for construction risk awareness. Those who were educated up to pre-university level, on the other hand, obtained a corresponding mean score of 59.68. This difference is statistically significant as shown by the p value of .01 in the independent samples t-test analysis conducted. This implies that university education helped improve awareness of construction risk.

## 16.6 Research on Risks Related to Construction

Although Malaysia has developed expertise and tools such as environmental auditing and environmental impact assessment to help quantify impacts and risks, it is hardly keeping up with the pace of development we are currently experiencing (Lee and Fong 2002). To date, only limited research has touched on how to integrate the identification, assessment, and management of construction-related risks to the environment. Measures to eliminate or assuage these risks must be properly identified and formulated at the outset of the project planning process to include engineering, geotechnical, and architectural inputs, among others (Eom and Paek 2009). Research approaches by Eom and Paek (2009), Tam et al. (2004), Shen et al. (2005), Li et al. (2005), Cheung et al. (2004), and Chen et al. (2005) can be reviewed and adopted where suitable. However they are not without drawbacks, because their construction risk evaluations are based on the availability of guidelines.

Other research arbitrarily and inadequately identifies environmental impacts without taking account of site selection and building location. This leads to inaccurate conclusions as a result of generalized, skewed judgments. Dione et al. (2005) conducted a survey of multiple construction companies, and the companies showed concern about the possible implications of environmental risks resulting from their projects. However, there still needs to be more emphasis on identifying and mitigating such risks.

In order to maintain control over the risks incurred from construction, it is imperative for aspects of construction that create these risks to be explicitly defined. Research to date is still inadequate to do this systematically, and it is also not yet possible to assess the level of risks or to prioritize measures to assuage these risks. An environmental management plan (EMP) is an integral part of assessing the impacts of a construction project, and yet during the monitoring of environmental matrices, many of the impacts and risks remain unmitigated (Saunders and Bailey 1999). Another research consideration is to link the practice of environmental impact assessment with environmental management plans, environmental auditing, and environmental monitoring to strengthen efforts to protect the environment from the insensitivities of construction (Nawrocka and Parker 2009).

## 16.7 Conclusion

The construction industry is characterized by a high degree of fragmentation, with numerous individual participants each pursuing singular interests on a project-by-project basis. Construction activities by their very nature cause disturbances or risks to the environment. However, these risks can be identified following simple procedures, and measures can be implemented to mitigate them. Little research has been conducted on the effect of construction on the environment yet there are steps and guidelines available to assist the industry in minimizing risks to the surroundings.

## References


Chen Z, Li H, Wong CTC (2005) Environmental planning: analytic network process model for environmentally conscious construction planning. J Constr Eng Manage 131(1):92–101, JCEMD40733–9364

Cheung SO, Tam CM, Tam V, Cheung K, Suen H (2004) A web-based performance assessment system for environmental protection: WePass. Constr Manage Econ 22(9):927–35, 0144-6193

Christini G, Fetsko M, Hendrickson C (2004) Environmental management systems and ISO 14001 certification for construction firms. J Constr Eng Manag 130(3):330–336

Dione S, Ruwanpura JY, Hettiaratchi JY (2005) Assessing and managing the potential environmental risks of construction projects. Pract Period Struct Des Constr 10(4):260–266, PPSCFX1084-0680

Eom CSJ, Paek JH (2009) Risk index model for minimizing environmental disputes in construction. J Constr Eng Manage 135(1):34–41

Gangolells M, Casals M, Gasso S, Forcada N, Roca X, Fuertes A (2011) Assessing concerns of interested parties when predicting the significance of environmental impacts related to the construction process of residential buildings. Build Environ 46:1023–1037

Hickie D, Wade M (1997) The development of environmental action plans: turning statements into actions. J Environ Plan Manag 40(6):789–801

Lee J, Fong TY (2002) Environmental management within the development approval process in the construction industry. In: Pereira JJ, Zainal Z, Hamid ZA (eds) Good environmental practices in the construction industry. Universiti Kebangsaan Malaysia, Bangi, pp 7–26

Li S, Shen Y, Zhang Z (2005) A new quantitative framework of environmental performance assessment applicable to construction stage. In: International conference on construction and real estate management. China Scientific Book Services, Beijing

Nawrocka D, Parker T (2009) Finding the connection: environmental management systems and environmental performance. J Clean Prod 17(6):601–607

Olander S, Landin A (2005) Evaluation of stakeholder influence in the implementation of construction projects. Int J Proj Manag 23:321–328

Põder T (2006) Evaluation of environmental aspects significance in ISO 14001. Environ Manag 37(5):732–743

Rodríguez G, Alegrea FJ, Martíneza G (2007) The contribution of environmental management systems to the management of construction and demolition waste: the case of the Autonomous Community of Madrid (Spain). Resour Conserv Recycl 50(3):334–349

Sanvicens GDE, Baldwin PJ (1996) Environmental monitoring and audit in Hong Kong. J Environ Plan Manag 39(3):429–440

Saunders AM, Bailey J (1999) Exploring the EIA/environmental management relationship. Environ Manag 24(3):281–295

Shen LY, Lu WS, Yao H, Wu DH (2005) A computer-based scoring method for measuring the environmental performance of construction activities. Autom Constr 14(3):297–309, AUCOES0926-5805

Tam VWY (2008) On the effectiveness in implementing a waste-management-plan method in construction. Waste Manag 28(6):1072–1080

Tam CM, Tam VWY, Tsui WS (2004) Green construction assessment for environmental management in the construction industry of Hong Kong. Int J Proj Manag 22(7):563–571, IPMAEL

Zobel T, Burman JO (2004) Factors of importance in identification and assessment of environmental aspects in an EMS context: experiences in Swedish organizations. J Clean Prod 12:13–27

Zobel T, Almroth C, Bresky J, Burman JO (2002) Identification and assessment of environmental aspects in an EMS context: an approach to a new reproducible method based on LCA methodology. J Clean Prod 10:381–396

# Chapter 17
# Ecosystem Restoration Using the Near-Natural Method in Shanghai

**Liang Jun Da and Xue Yan Guo**

**Abstract** Recently, there has been a growing trend toward developing eco-cities that are rationally structured, function efficiently, and maintain a harmonious relationship with the environment. Functional eco-cities demonstrating symbiosis with nature are not merely a current urban development goal, but also make an important contribution to the establishment of sustainable development strategies across the world. The "near-natural" method of ecological construction and restoration has attracted attention worldwide, and has been widely proven to be an effective method of constructing ecological cities in practice. It will also play a key role in urban landscaping and greening in China. After studying the problems associated with constructing an eco-city in Shanghai, we proposed the use of the near-natural method to construct an urban ecosystem in the city and introduced a theory and methodology for creation of "near-natural forests" and "near-natural water systems." All the cases we studied demonstrated that restoration using the near-natural method could be more effective, long lasting, and economical than existing methods, and that the method is now worthy of promotion as a means of constructing environments for human settlement.

**Keywords** Near-natural • Ecological restoration • Urban greening • Native species

L.J. Da (✉)
Department of Environmental Science, East China Normal University,
500 Dongchuan Rd., Shanghai 200241, China

Shanghai Key Lab for Urban Ecological Processes and Eco-Restoration, East China Normal University, 500 Dongchuan Rd., Shanghai 200241, China

Tiantong National Station of Forest Ecosystem, East China Normal University,
Normal University, Ningbo, Zhejiang, Shanghai 315114, China
e-mail: ljda@des.ecnu.edu.cn

X.Y. Guo
Department of Environmental Science, East China Normal University,
500 Dongchuan Rd., Shanghai 200241, China
e-mail: xueer308@sina.com

N. Kaneko et al. (eds.), *Sustainable Living with Environmental Risks*,
DOI 10.1007/978-4-431-54804-1_17, 

## 17.1 Introduction

The city, a typical artificial system constructed according to the will of humans, is an ecosystem composed of social, economic, and natural sub-ecosystems. The natural sub-ecosystem suffers drastic damage as a result of long-term human activities and rapid urbanization. The latter in particular causes many ecological problems, such as loss of native species and invasion of alien species, due to environmental pollution (Cao et al. 2008; Kan et al. 2004; Ren et al. 2003; Wang et al. 2008; Zhang et al. 2010) and local climate change (Xu et al. 2008; Zhang et al. 2009). This has a significant impact on the city's survival and development.

As a large metropolis characterized by a high-density population, concentrated industries, and resource shortages, Shanghai has also experienced such problems. Consequently, its leaders now realize the importance of healthy ecosystems to the city's process of development into an eco-city in the new century. In recent years, Shanghai has made great progress in restoring the natural sub-ecosystem, but some problems still remain due to the desire to achieve its aims quickly using simple methods to shorten the restoration process.

One example is employing monoculture plantations—usually consisting of conifers or other rapidly growing pioneer trees—to restore degraded terrestrial landscape. Monoculture plantations are, however, quite vulnerable to plant diseases and insect pests (Gibson and Jones 1977; Ciesla and Donaubauer 1994; Gadgil and Bain 1999) due to their extreme homogeneity. Moreover, high-density planting of these monotypic stands limits the resources available, inducing intense competition and low seedling and sapling recruitment rates. As a consequence, these forests always exhibit simple vertical structure and are unstable, in addition to harboring a lower animal diversity (Kloor 2000).

A second example is using a large proportion of non-native species in urban greening for decorative and aesthetic reasons, while natural vegetation comprising native species suitable for local habitats is rarely used. Most non-native species are not well suited to their location, which results in low survival rates and high management costs. As a result, the sustainability of greening dominated by non-native species is low and its ecological function weak.

Third, water area—which has shrunk greatly during recent decades—is often increased using water systems devoid of plant or animal life, or waterscapes with concrete bottoms and revetments. Curved stream channels are also straightened to prevent or discharge floods, and concrete revetments have replaced the natural littoral zone. As a result, all such water areas have lost the connection between water and soil and cannot provide habitats for living beings, resulting in weak self-purification capacity and accumulation of nutrients in the water. Algae bloom occurs frequently without intact aquatic ecosystems and hierarchical trophic webs (Kondolf 1995), especially as a result of losing submerged vegetation and herbivores.

Hence, if the negative consequences of the above three methods are to be avoided, it is important to comply with ecological principles to restore the structure and function of the degraded urban natural sub-ecosystem, both terrestrial and aquatic

(Hobbs 1996; Allen et al. 1997; Palmer et al. 1997). This remains a pressing goal for urban areas (Kloor 2000). Based on experience gained by many researchers over several decades (Miyawaki 1998; Wang et al. 2002), we used a method emulating natural processes—specifically the "near-natural" method as described below—to restore the degraded natural sub-ecosystem in Shanghai.

## 17.2 Theoretical Basis for Near-Natural Restoration

### *17.2.1 The Near-Natural Method of Afforestation*

The near-natural method is based on the concepts of potential natural vegetation and succession theory in vegetation ecology (Miyawaki 1998; Wang et al. 2002). In natural forest ecosystems, the succession from the establishment of pioneer plant communities to forested climax communities can take hundreds of years. However, the near-natural approach, using seedlings of native woody species raised in plastic pots to revegetate reformed soil, can accelerate succession and create a near-climax community in a shorter time period and at low cost. This near-climax community would have a complex structure with high species diversity, multilayer canopies, and high biomass in a stable state, requiring less management.

### *17.2.2 The Near-Natural Method of River Construction*

In 1938, Seifert raised the concept of near-natural river construction (Seifert 1938). In the last two decades, the popularity of ecological concepts in Japan has led to the construction of attractive waterside spaces, recovery and creation of living rivers, and improvements in river environments and ecosystems (Zhu et al. 2006). Compared with traditional projects, the prominent characteristic of near-natural construction is that biological diversity, density, and productivity increase several times over (Gao 1999).

## 17.3 Near-Natural Forest Creation

A 3,000 $m^2$ near-natural forest was established in the Pudong New Area of Shanghai (31°13′N, 121°32′E) in June 2000. The climate of this region is subtropical monsoon with a mean annual temperature of 18.1 °C and precipitation of 1,158 mm. The potential natural vegetation local to Shanghai is dominated by evergreen broad-leaved forest like *Cyclobalanopsis glauca, C. myrsinaefolia, Castanopsis sclerophylla*, and *Machilus thunbergii*. Due to the high water table and high soil salinity

**Table 17.1** Species list of planted seedlings for a near-natural forest in Pudong, Shanghai, in 2000. All species are native to the local region, where evergreen broad-leaved forest is considered to be the zonal vegetation

| Plant species | Life form | Density(per 3,000 m$^2$) |
|---|---|---|
| *Cyclobalanopsis myrsinaefolia* | Evergreen tree | 2,500 |
| *Cyclobalanopsis glauca* | Evergreen tree | 2,500 |
| *Castanopsis sclerophylla* | Evergreen tree | 2,500 |
| *Machilus thunbergii* | Evergreen shrub | 2,830 |
| *Ligustrum lucidum* | Evergreen shrub | 370 |
| *Distylium racemosum* | Evergreen shrub | 150 |
| *Pittosporum tobira* | Evergreen shrub | 400 |
| *Fatsia japonica* | Evergreen shrub | 300 |
| *Aucuba chinensis* var.*variegata* | Evergreen shrub | 300 |
| *Liquidambar formosana* | Deciduous tree | 300 |
| *Swida alba* | Deciduous shrub | 200 |
| Total | | 12,350 |

in Shanghai, flat ground was reformed into a landscape with micro-topographies to create suitable habitats for climax woody species before transplanting seedlings.

All species of seedling used in the afforestation were native to Shanghai and surrounding areas (Table 17.1). Seedlings were raised in greenhouses until they developed a strong root system and reached a height of 0.2 to 0.8 m. They were planted at densities as high as three to five individuals per m$^2$, because high-density planting favors natural selection of the best seedlings that match their microsite conditions, at the same time providing a source of additional seedlings to meet greening needs in the future. After planting, the bare ground was covered with rice straw and weeding was undertaken every 2 or 3 years. After 5 years, no further management was necessary.

### *17.3.1 Monitoring: Stand Dynamics and Eco-Benefits*

Two adjacent plots, each of 200 m$^2$, were set up. In each plot, the height and diameter at breast height for each individual taller than 1.5 m were measured in 2001, 2003, 2004, 2006, and 2010.

To assess the ecological benefits of the near-natural forest, the forest was compared with an evergreen broad-leaved forest in Ningbo, near Shanghai, and other artificial forests in Shanghai (Table 17.2), based on measurements of negative air ions (NAI), air temperature and humidity, air microorganisms, and soil microorganisms in 2007.

NAI, and air temperature and humidity, were measured in the center of each plot at a height of 1.5 m above ground every 2 h between 9:00 and 17:00 over three consecutive days in May. Measurements were recorded for 10 min at a time using an NAI monitor (ITC-201A), which can measure second-by-second and automatically save sets of data.

**Table 17.2** To assess the ecological benefits, the near-natural forest in Pudong, Shanghai, was compared with natural mature forests, artificial forests, and other non-forest sites, such as lawn and bare ground, in 2007

| Plots | Vegetation | Location | Dominant species | Maximum height (m) | Canopy cover (%) | Concentration of air anions, air temperature and humidity | Air microorganisms | Soil microorganisms |
|---|---|---|---|---|---|---|---|---|
| NF1 | Natural forest | Tiantong,Ningbo | *Castanopsis fargesii, Schima superba* | 20 | 85 | √ | | √ |
| NF2 | Natural forest | Tiantong,Ningbo | *Castanopsis fargesii, Schima superba* | 18 | 85 | √ | | √ |
| NN1 | Near-natural forest | Pudong | *Cyclobalanopsis glauca, Machilus thunbergii, Cyclobalanopsis myrsinaefolia* | 12 | 95 | √ | √ | √ |
| NN2 | Near-natural forest | Pudong | *Cyclobalanopsis glauca, Machilus thunbergii, Cyclobalanopsis myrsinaefolia* | 12 | 90 | | √ | √ |
| AF1 | Artificial forests | Changfeng park | *Magnolia grandiflora, Magnolia denudate* | 9 | 90 | √ | | √ |
| AF2 | Artificial forests | Pudong | *Cinnamomum camphora, Photinia serrulata, Fatsia japonica, Oxalis corniculata* | 10 | 80 | √ | √ | √ |
| AF3 | Artificial forests | Pudong | *Magnolia grandiflora, Cinnamomum camphora, Ophiopogon japonicus* | 10 | 50 | √ | | |
| CG | Lawn | East China Normal University(ECNU) | *Cynodon dactylon* | 0.1 | – | √ | | |
| CK | Bare ground | ECNU | – | – | – | | √ | |

Because there are no natural mature forests in Shanghai due to long-term human disturbance, two natural forests in a national forest park were used. They are located close to Shanghai in Ningbo, and belong to the same vegetation zone as Shanghai. The other sites were located in urban areas of Shanghai and their dominant species were the most common greening species in Shanghai. The artificial forests also had similar vegetation structure to the near-natural forest. Three categories of indicator were measured to evaluate ecological benefits. The √ mark denotes indicators measured

Air microorganisms were collected using an air microorganism sampler (FA-1) at a height of 1.5 m above ground for 10 min, with three replications in each plot in spring, summer, fall, and winter, respectively. The culture media for air bacteria, air fungi, and air actinomycetes were the nutrient agar, Rose Red sodium agar, and Gause No.1, respectively. After sampling, the media were taken back to the laboratory to culture and count.

Soil microorganisms were cultured by means of the spread plate method using similar media to those used for air microorganisms. The soil samples were obtained from each plot in May at five points distributed in a "Z" shape at 0–10 cm and 10–20 cm depths. These samples were mixed using the multi-point soil samples method, taken back to the laboratory and diluted for determination.

## 17.3.2 Monitoring Results

### 17.3.2.1 Plant Species Composition and Dynamics

The seedling survival rate the first winter was greater than 90 %. From 2001 to 2010, the species composition remained fairly stable, with the exception of the invasion of several native non-planted species (Table 17.3).

Until 2010, the tenth year after transplanting, the average heights of *Liquidambar formosana* and *Ligustrum lucidum* were 9.2 m and 6.5 m, respectively (Fig. 17.1). The average heights of *Machilus thunbergii*, *Cyclobalanopsis glauca*, and *Distylium racemosum* were lower in 2010 than in 2006, because of the death of several larger individuals and the recruitment of considerable numbers of seedlings. In addition, native deciduous species, including *Ulmus parvifolia*, *Sapium sebiferum*, *Salix matsudana*, *Broussonetia papyrifera*, *Cinnamomum camphora*, and *Podocarpus macrophyllus*, invaded the stand, via seed dispersal by wind and by birds. In 2010, the largest individual was *Sapium sebiferum*, at 12.5 m in height, while *Broussonetia papyrifera* , the early invader, had reached a height of 9.5 m.

*Ligustrum lucidum* and *Liquidambar formosana* always topped the list of average breast-height basal areas from 2003 onwards (Table 17.3). In addition, it is notable that the breast-height basal areas of *Cyclobalanopsis myrsinaefolia* and *Cyclobalanopsis glauca* were smaller in 2004 than in 2003. Similarly, those of the *Machilus thunbergii* and *Ligustrum quihoui* were smaller in 2010 than in 2006, as a result of the growth of considerable numbers of new seedlings into the shrub layer and the death of several larger individuals.

### 17.3.2.2 Ecological Benefits

The concentrations of NAI in the near-natural forest were much higher than in artificial forests in urban areas, but much lower than in true natural forests. This is because high tree density and a multi-layer vertical structure can effectively increase

**Table 17.3** Variation of species composition, density (ind/3,000 m$^2$) and basal area at height of 1.5 m (BA, cm$^2$/3,000 m$^2$) in the near-natural forest in Pudong, Shanghai from 2001 to 2010. **Cerasus serrulata* var. *lannesiana* was planted in 2003

| | | 2001 | 2003 | | 2004 | | 2006 | | 2010 | |
|---|---|---|---|---|---|---|---|---|---|---|
| | Species | Density | Density | BA | Density | BA | Density | BA | Density | BA |
| Artificial planted species | *Cyclobalanopsis myrsinaefolia* | 182 | 143 | 288.4 | 142 | 242.8 | 144 | 560.3 | 109 | 439.9 |
| | *Cyclobalanopsis glauca* | 81 | 73 | 171.4 | 69 | 152.3 | 70 | 352.4 | 68 | 302.6 |
| | *Castanopsis sclerophylla* | 169 | 107 | 15.6 | 35 | 10.3 | 12 | 27.5 | 5 | 83.4 |
| | *Machilus thunbergii* | 68 | 45 | 23.8 | 44 | 54.8 | 35 | 89 | 29 | 36.3 |
| | *Ligustrum lucidum* | 27 | 24 | 572.9 | 28 | 1,206 | 28 | 1,860 | 21 | 1,002.54 |
| | *Distylium racemosum* | 12 | 9 | 9.8 | 11 | 16.4 | 30 | 36.7 | 38 | 30.2 |
| | *Pittosporum tobira* | 36 | 28 | | 34 | 215.9 | 43 | 442.5 | 31 | 473.1 |
| | *Fatsia japonica* | 11 | 7 | | 8 | | 10 | | 6 | |
| | *Aucuba chinensis* var.*variegata* | 1 | 1 | | 1 | | 1 | | 1 | |
| | *Liquidambar formosana* | 28 | 28 | 164.6 | 26 | 287.0 | 27 | 666.6 | 20 | 1,485.1 |
| | *Swida alba* | 2 | 1 | | 0 | | 0 | | – | |
| | *Cerasus serrulata* var. *lannesiana** | – | 4 | | 4 | | 1 | | 2 | |
| Natural invading species | *Cinnamomum camphora* | 1 | 1 | | 3 | | 9 | | 4 | |
| | *Sapium sebiferum* | 1 | 1 | | 1 | | 1 | | 1 | |
| | *Ulmus parvifolia* | 1 | 1 | | 1 | | – | | – | |
| | *Salix matsudana* | 1 | – | | – | | – | | – | |
| | *Broussonetia papyrifera* | – | 3 | | 1 | | 69 | | 43 | |
| | *Podocarpus macrophyllus* | – | – | | – | | 1 | | 4 | |
| | *Trachycarpus fortunei* | – | – | | – | | – | | 11 | |
| | *Celtis sinensis* | – | – | | – | | – | | 1 | |
| Total | | 621 | 476 | | 408 | | 481 | | 394 | |

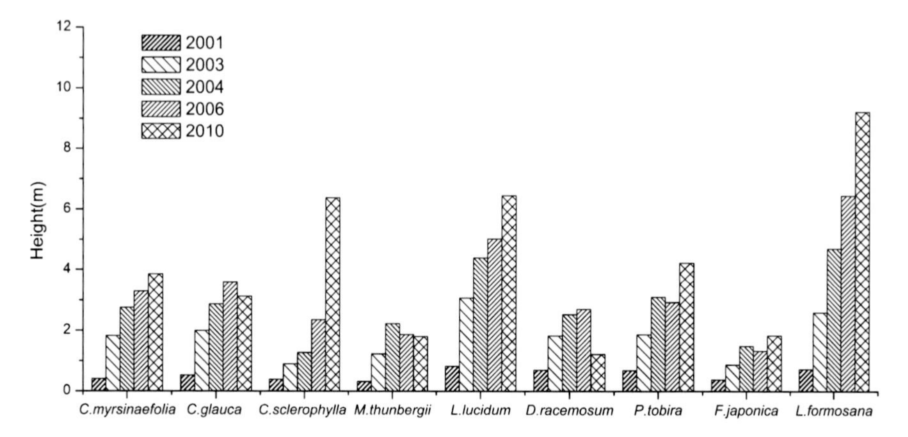


**Fig. 17.1** Dynamics of the main species in terms of average height

the concentration of NAI (Shao and He 2000). Air temperature and humidity, meanwhile, demonstrated no significant difference between the near-natural forest and artificial forests in urban areas. The average air bacteria concentration of the near-natural forest was lower than the Pudong and East China Normal University sample plots, but air fungi and actinomycetes were higher. The near-natural forest's air bacteria concentration was also lower than that of artificial forests, while air fungi and actinomycetes were again higher. Within the soil, meanwhile, the abundance of microorganisms in the near-natural forest was greater than in artificial forests, and close to the abundance in true natural forests, due to thicker litter in near-natural forest which can increase soil microorganisms to improve soil fertility (Table 17.4) (Alexander 1983; Huang and Wu 2002).

In this area, a near-natural forest had developed within 10 years. Since both canopy and subcanopy species were planted, the forest exhibited a vertical structure consisting of layers of trees, shrubs, and herbaceous plants. This vertical structure fosters greater species diversity by providing more niches for organisms. The area has now grown into beautiful woodland with remarkable effects on the landscape, and has provided considerable ecological benefits (Fig. 17.2).

### *17.3.3 Assessment of Near-Natural Forest Experiment and Proposal for Improvement*

Because the near-natural forest experiment used native species and involved little management, the costs were limited considerably. The forest ultimately consists of both the planted native species and those species that invaded naturally. Diversification of species has therefore continued without any additional cost. The variety of species will enhance resistance to pests, diseases, and other disturbances.

**Table 17.4** Comparison of ecological benefits among near-natural forest in Pudong, natural mature forests, artificial forests, and lawn. Results are shown using mean ± SD. A One-Way ANOVA test was used to check the significance

| | | | | Air | | | Soil | | |
|---|---|---|---|---|---|---|---|---|---|
| Plot | Air temperature (°C) | Air humidity (%) | Negative ions (ind/cm$^3$) | Bacteria (ind/m$^3$) | Fungi (ind/m$^3$) | Actinomycetes (ind/m$^3$) | Bacteria (ind/g) | Fungi (ind/g) | Actinomycetes (ind/g) |
| NF1 | 15.3 ± 1.1a | 83.3 ± 10.2a | 954 ± 88.9a | | | | 1.6E+08a | 5.7E+05a | 7.7E+05a |
| NF2 | 17.0 ± 0.2a | 47.5 ± 9.2b | 1,534 ± 140.8b | | | | 1.8E+07ab | 1.3E+05a | 5.7E+05a |
| NN1 | 22.8 ± 1.7b | 60.6 ± 9.4b | 420 ± 63.7c | 818 ± 148a | 839 ± 843ab | 36 ± 9a | 4.2E+08ab | 2.0E+05b | 3.1E+06a |
| NN2 | | | | 896 ± 151a | 1,017 ± 623b | 26 ± 17b | 3.9E+09ab | 1.9E+05b | 2.6E+06b |
| AF1 | 22.3 ± 2.0b | 57.1 ± 9.0b | 340 ± 29.7d | | | | 4.8E+10c | 1.4E+05c | 2.8E+06ab |
| AF2 | 22.7 ± 1.6b | 58.0 ± 9.2b | 280 ± 27.3de | 1,008 ± 171b | 758 ± 395c | 19 ± 13c | 8.8E+10bc | 1.4E+05bc | 3.0E+06b |
| AF3 | 24.5 ± 1.5b | 59.8 ± 9.3b | 220 ± 97.3e | | | | | | |
| CG | 24.5 ± 2.6b | 51.6 ± 10.4b | 90 ± 53.5f | | | | | | |
| CK | | | | 2,225 ± 1,496c | 366 ± 259d | 10 ± 5d | | | |

Different letter in the same column means significant difference ($p < 0.05$)

**Fig. 17.2** (**a**) Initial planting (June 11, 2000); (**b**) Same site, four years later (June 8, 2004); (**c**) Seven years later (July 4, 2007)

Moreover, near-natural forests require much less maintenance than other forms of urban greening, particularly in the later stages of growth.

Due to its structure and the nature of its ecological benefits, the near-natural forest formed a stable community after 10 years of development. It also provided ecological benefits on a greater scale than traditional artificial forests, rendering it much more similar to natural forests. In addition, many birds and insects were recorded in the near-natural forest, indicating a potential role as a refuge for animals in urban settings. We consider that further increases in ecological benefits are likely to occur as the near-natural forest grows, and foliage increases.

But one notable problem during this experiment was the high mortality rate among several climax evergreen broad-leaved species in the first few years. These were species intended as the dominant component of the near-natural forest at the mature stage. Investigation revealed that the mortality rate was related to the lack of shade to protect seedlings against overexposure to strong sunlight. This conclusion was supported by the fact that *Cyclobalanopsis myrsinaefolia* and *Cyclobalanopsis glauca* demonstrated higher survival rates in 2004 due to the shade provided by fast-growing *Ligustrum lucidum* and *Liquidambar formosana*.

To further optimize the method for near-natural forest creation in Shanghai, we planted another two near-natural forests during the winter of 2003 in the city's Minghang district. The seedlings of the target climax community demonstrated a higher growth rate with lower mortality than in the original experiment. In this case, the deciduous trees grew faster than the evergreen ones, and the resulting closed canopy shielded the evergreen target seedlings from overexposure. To optimize the traditional Miyawaki Method (Miyawaki 1998; Miyawaki 1999), therefore, we proposed a planting method based on multilayer deciduous-evergreen mixed forests comprising trees of different ages. The key to the new method is to create a mixed deciduous-evergreen community by simultaneously planting shade-tolerant evergreen broad-leaved species and light-demanding deciduous broad-leaved species, but using smaller individuals for the former and bigger individuals for the latter to form a multilayer vegetation structure. The shade-tolerant evergreen species benefit from the rapid growth of the light-demanding deciduous species, which offer shade and nutrients in the form of litter layer-based fertilizer, improving the soil for the evergreen species.

## 17.4 Near-Natural River Construction

Degradation of aquatic ecosystems caused by eutrophication is a common problem in China's rivers and lakes in recent years. Liwa River passes through the campus of East China Normal University, Shanghai (Fig. 17.3), and its water quality was previously so poor that it failed to qualify for Class V (the least stringent category) under China's National Surface Water Environmental Quality Standard (GB3838-2002). Its nitrogen and phosphorus concentrations were two to four times higher than the required standard. Comprehensive treatment of Liwa River was initiated in August 2004. By the end of 2005, rehabilitation was complete, including

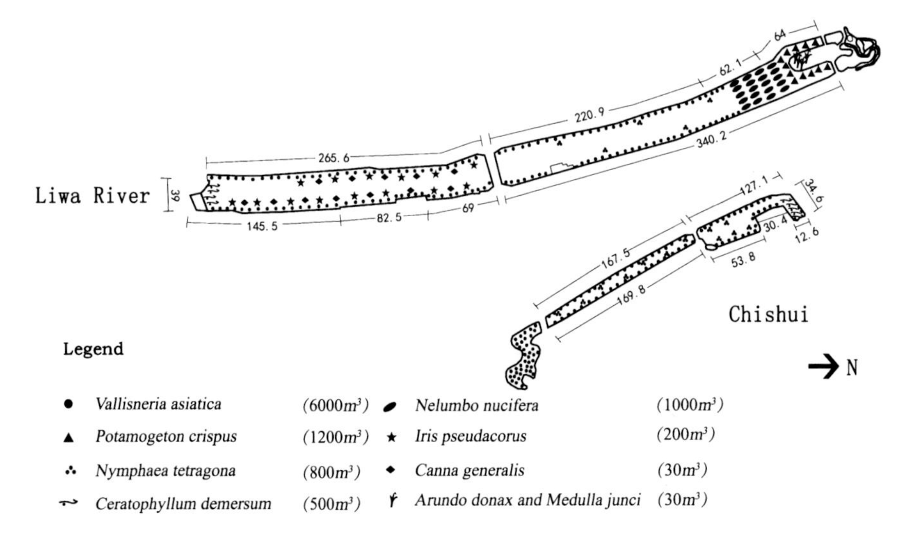


**Fig. 17.3** Aquatic plant growing sites in Liwa River (Unit: m)

reducing pollutant input, dredging, improving water cycling, and setting up a membrane bioreactor. Near-natural restoration work to establish an aquatic ecosystem with an integrated trophic web started in April 2006, and ended in November the same year. The ecosystem was created using three ecological engineering systems based on sediment dredging and prevention of pollution.

1. Engineering of Water Quality Control
   Methods of river-base solidification and biology were used to increase water transparency and ameliorate water quality. Fountains were used to increase the concentration of dissolved oxygen and change the grade distribution, which can enhance capacity for decomposition of organic material and remove excess nutrients from water.
2. Engineering of Water Landscape Construction
   To naturalize the concrete banks along Liwa River without any destruction, plant species with trailing branches, such as *Jasminum mesnyi* and *Hedera nepalensis var. sinensis*, were used to cover the banks, while emerging plants on fiber drums floating on the water near the shoreline alleviate erosion and enhance the sense of a visual landscape.

   Since aquatic vegetation is highly effective in removing pollutants, from April to June 2006 macrophyte communities dominated by native species were reestablished to increase the area of aquatic vegetation and biomass in the river. The aquatic vegetation comprised submerged plants, emerging plants, and floating plants, to form a three-layer vertical community structure. Hydrophyte, helophyte, hygrophyte, and mesophyte species were selected for reestablishment, some of which offered efficient nutrient removal or allelopathic effects on algae (Li et al. 2006; Chen et al. 2009). In addition, floating islands made of fiber mats were set up in areas where the depth was beyond the recruit limit of aquatic vegetation, and floating-leaved plants such as *Nymphaea tetragona* and emerging plants were planted on the mats.
3. Construction of Food Web in Water and Engineering of Biological Diversity
   After aquatic vegetation was established, planktivores (*Hypophthalmichthys molitrix* and *Aristichthys novilis*) and an invertebrate (*Bellamya quadrata*) were introduced to the river in November 2006. *B. quadrata* was introduced at between 0.5 and 1.0 kg/m$^3$, with adult individuals dominating. *H. molitrix* and *A. novilis* were introduced at 0.5–0.6 kg/m$^3$ (about 2.0 fish per m$^3$) and 0.05–0.06 kg/m$^3$ (about 0.2 fish per m$^3$), respectively, with young individuals dominating. These species are not only part of the local food web, but are also powerful inhibitors of algal bloom (Starling and Rocha 1990).

   Subsequently, *Carassius auratus*, *Parabramis pekinensis* and other native waterfowl at higher trophic levels were added into the river to form an ecosystem with a long web chain including producer, herbivore, predator, and decomposer. Eventually, indirect export of organic matter and nutrients from the water could be achieved by fishing, capturing aquatic animals, and harvesting plants.

## *17.4.1 Monitoring: Water Quality and Biocoenosis*

### 17.4.1.1 Water Quality

Eight water samples were obtained 0.5 m underwater and determined according to the National Surface Water Environmental Quality Standard (GB3838-2002). The parameters were measured as follows: total phosphorus (TP) using the ammonium molybdate spectrophotometric method; total nitrogen (TN) via the alkaline potassium persulfate digestion-UV spectrophotometric method; chemical oxygen demand (COD) by the dichromate method; dissolved oxygen (DO) using the iodimetric method; Chlorophyll-a via spectrophotometry; and water transparency by Secchi disc.

### 17.4.1.2 Aquatic Plants

The fresh weight was measured with three replications by harvesting all plants in a square of 1 $m^2$ in each season. After that, fresh samples were air-dried, smashed and sieved for later determination. Total phosphorus and total nitrogen were measured using the perchloric acid-sulfuric acid solution method.

### 17.4.1.3 Aquatic Animals

Samples of zoobenthos were collected by 1/16 $m^2$ Peterson grab at three points. Samples of fish were obtained by fishing in a certain location. The fresh weight was measured after washing and identification.

## *17.4.2 Monitoring Results*

### 17.4.2.1 Variation of Water Quality

In general, water quality was improved from below Class V to Class II based on the National Surface Water Environmental Quality Standard (GB3838-2002) (Table 17.5). TN, TP, COD, and Chlorophyll-a were greatly decreased. From 2003 to 2007, the transparency increased from 20–40 to 80–100 cm. The DO was low with an average value less than 5 mg/L and a minimum of 0.8 mg/L in summer 2003, increasing to 9.09–11.18 mg/L in November 2007. The concentration of Chlorophyll-a decreased from an average level of 19.41 μg/L to around 7.86 μg/L. Algae was remarkably inhibited after introducing *H. molitrix* and *A. novilis*. In addition, inhibition of algal growth via competition from aquatic plants was immeasurable (Chen et al. 2009).

**Table 17.5** Water quality of Liwa River

| Date | Engineering phase | TN (mg/L) | TP (mg/L) | COD (mg/L) | Chlorophyll-a (μg/L) |
|---|---|---|---|---|---|
| 2003.10 | Before treatment | 11.34 | 1.13 | 58.9 | 123.52 |
| 2005.11 | Traditional treatment completed | 1.85 | 0.23 | 33.3 | – |
| 2006.11 | Near-natural restoration completed | 0.89 | 0.19 | 20.2 | 19.41 |
| 2007.11 | One year later | 0.41 | 0.03 | 21.5 | 7.86 |

#### 17.4.2.2 Biocoenotic Dynamics

The distribution and biomass of each introduced species was recorded monthly for 2 years after restoration (Table 17.6). In comparison with their areas in 2006, the ranges of *P. crispus*, *N. nucifera*, and *Ceratophyllum demersum* expanded 96 %, 100 %, and 84 %, respectively. In contrast, *Vallisneria asiatica* and *Iris tectorum* lost 72 % and 50 % of their ranges, respectively. Range shrinkage of *V. asiatica* was observed only in deep-water areas, which suggests that the decrease in *V. asiatica* was mainly due to insufficient light (Li et al. 2008).

Most aquatic species, such as *V. asiatica* and *C. demersum*, attained maximum biomass in summer and fall, except for *P. crispus*. Therefore, *P. crispus* played an important role in removing nutrients, filtering particles, and replenishing oxygen in winter when the other species lost vitality. As a result, the aquatic vegetation had an observable effect on the removal of excess nutrients throughout the whole year and provided habitats and sufficient food for fish even in the winter.

Most aquatic plants reached the highest fix rate of nitrogen and phosphorus in fall, but *P. crispus* and *N. nucifera* were the highest in spring and in summer respectively. Calculations based on data collected showed that 12.8 kg of N and 1.6 kg of P could be removed from the water by harvesting *P. crispus* in May and *N. nucifera* in November every year (Table 17.7). The effect would be even better if *N. nucifera* was harvested in summer. The data indicated that harvesting aquatic vegetation was an effective treatment for urban river eutrophication (Graneli and Solander 1988; Meuleman et al. 2004).

Two years after the restoration, native zoobenthos species *Radix swinhoei* and *Planorbidae spp*, and native fish *Saurogobio dumerili* were found in addition to the species artificially introduced. Invertebrates in particular were thriving in the flourishing macrophyte communities. The number of *B. quadrata*, *R. swinhoei* and *Planorbidae spp* increased to 82.5, 151.5, and 19.3 $m^3$, respectively. These increases showed that food chains had reestablished with increased species richness in the restored river.

**Table 17.6** Distribution and biomass of aquatic plants

| Life form | Species | Areas in 2006 ($m^2$) | Proportion in 2006 (%) | Areas in 2008 ($m^2$) | Proportion in 2008 (%) | Total fresh weight (kg) | | | |
|---|---|---|---|---|---|---|---|---|---|
| | | | | | | Spring | Summer | Fall | Winter |
| Submerged plants | *Potamogeton crispus* | 1,200 | 4.1 | 2,350 | 8 | 1,913 | – | 120 | 672 |
| | *Vallisneria asiatica* | 6,000 | 20.5 | 1,660 | 5.7 | 883 | 1,117 | 905 | 654 |
| | *Ceratophyllum demersum* | 500 | 1.7 | 920 | 3.1 | 244 | 458 | 289 | 222 |
| Emerging plants | *Nelumbo nucifera* | 1,000 | 3.4 | 2,000 | 6.8 | – | 2,250 | 650 | – |
| | *Medulla junci* | 30 | 0.1 | 30 | 0.1 | 16 | 40 | 34 | 12 |
| | *Arundo donax* | 30 | 0.1 | 30 | 0.1 | 17 | 45 | 37 | 13 |
| | *Canna generalis* | 30 | 0.1 | 30 | 0.1 | 6 | 20 | 16 | 12 |
| | *Iris pseudacorus* | 200 | 0.7 | 100 | 0.3 | 14 | 61 | 47 | 33 |
| Floating plants | *Nymphaea tetragona* | 800 | 2.7 | 760 | 2.6 | 151 | 314 | 249 | 107 |

**Table 17.7** Nitrogen and phosphorus fixed by aquatic plants

| | | Submerged plants | | | Emerging plants | | | | | Floating plants |
|---|---|---|---|---|---|---|---|---|---|---|
| Seasons | | *P. crispus* | *V. asiatica* | *C. demersum* | *N. nucifera* | *M. junci* | *A. donax* | *C. generalis* | *I. pseudacorus* | *N. tetragona* |
| Areas in 2008 ($m^2$) | Spring | 2,350 | 1,660 | 700 | – | 30 | 30 | 30 | 90 | 660 |
| | Summer | – | 1,660 | 920 | 2,000 | 30 | 30 | 30 | 100 | 800 |
| | Fall | 800 | 1,660 | 820 | 1,000 | 30 | 30 | 30 | 100 | 760 |
| | Winter | 1,640 | 1,660 | 650 | – | 30 | 30 | 30 | 90 | 500 |
| Amount Of Nitrogen ($g/m^2$) | Spring | 2.214 | 0.061 | 0.020 | – | 2.056 | 2.212 | 0.452 | 0.420 | 0.010 |
| | Summer | – | 0.600 | 0.380 | 10.356 | 9.780 | 10.210 | 4.760 | 4.612 | 0.190 |
| | Fall | 0.059 | 1.600 | 1.061 | 12.046 | 25.130 | 29.400 | 13.613 | 13.027 | 0.531 |
| | Winter | 0.267 | 0.243 | 0.154 | – | 3.430 | 3.660 | 4.040 | 3.757 | 0.077 |
| Amount Of Phosphorus ($g/m^2$) | Spring | 0.204 | 0.016 | 0.010 | – | 0.672 | 0.820 | 0.395 | 0.273 | 0.005 |
| | Summer | – | 0.151 | 0.101 | 1.520 | 1.346 | 2.520 | 1.524 | 1.248 | 0.051 |
| | Fall | 0.005 | 0.399 | 0.266 | 1.130 | 3.810 | 4.540 | 2.687 | 2.373 | 0.143 |
| | Winter | 0.025 | 0.061 | 0.040 | – | 1.202 | 1.360 | 1.322 | 1.180 | 0.020 |
| Total amount of Nitrogen (kg) | Spring | 4.990 | 0.101 | 0.014 | – | 0.062 | 0.066 | 0.014 | 0.038 | 0.007 |
| | Summer | – | 0.996 | 0.350 | 11.650 | 0.293 | 0.306 | 0.143 | 0.461 | 0.152 |
| | Fall | 0.047 | 2.656 | 0.870 | 7.830 | 0.754 | 0.882 | 0.408 | 1.303 | 0.403 |
| | Winter | 0.439 | 0.403 | 0.100 | – | 0.103 | 0.110 | 0.121 | 0.338 | 0.039 |
| Total amount of Phosphorus (kg) | Spring | 0.480 | 0.027 | 0.007 | – | 0.020 | 0.025 | 0.012 | 0.025 | 0.003 |
| | Summer | – | 0.251 | 0.093 | 3.040 | 0.040 | 0.076 | 0.046 | 0.125 | 0.041 |
| | Fall | 0.004 | 0.662 | 0.218 | 1.130 | 0.114 | 0.136 | 0.081 | 0.237 | 0.109 |
| | Winter | 0.041 | 0.101 | 0.026 | – | 0.036 | 0.041 | 0.040 | 0.106 | 0.010 |

**Table 17.8** Comparison of expenses for removing nitrogen and phosphorus between traditional treatment and near-natural restoration engineering

| | Total amount of N (kg) removed | Total amount of P (kg) removed | Total cost (million RMB) | Unit price of removing N (million RMB/kg) | Unit price of removing P (million RMB/kg) |
|---|---|---|---|---|---|
| Traditional treatment | 419.9 | 39.8 | 541.8 | 1.3 | 13.6 |
| Near-natural restoration | 63.7 | 8.9 | 19.9 | 0.3 | 2.2 |

### *17.4.3 Assessment of Near-Natural River Construction*

The total cost of comprehensive treatment in Liwa River was RMB 5.617 million. The cost of near-natural restoration was RMB 0.199 million, accounting for only 3.5 % of the total cost (Table 17.8).

There were only a few additional expenses for post-restoration management involving macrophyte vegetation harvesting every half year. Other methods typically cost four to six times more than near-natural restoration to remove the same amount of nutrients. This is because near-natural restoration almost exclusively uses solar energy, a very diffuse but sustainable and free energy source (Seidel 1976). Photosynthesis and renewable energy utilization are economical by nature. But it cannot be doubted that it would be a cheaper and more effective treatment for seriously polluted rivers to employ near-natural restoration after traditional environmental engineering techniques that are considered more likely to have an immediate effect.

## 17.5 Concluding Remarks

Near-natural restoration is both a method and a theory. Its meaning surpasses ecology, embracing a philosophy of "harmony between man and nature." Since the era of the Warring States (720–221 BC), the Chinese have investigated the harmonious relationship among Tian (heaven or universe), Di (earth or resource), and Ren (people or society), advocating the union of man and nature. From these investigations, a systematic set of principles for managing the relationship between man and the environment was developed. In particular, principles of holism, symbiosis, circulation and self-reliance were emphasized (Wong and Bradshaw 2008).

In successive phases, appropriate native species were selected, relevant communities from the pioneer to senior phase were built, and each phase of the food cycle was connected, to enable the forest or water system to become a self-maintaining, recycling, living ecosystem. Numerous restoration projects and studies have shown that near-natural restoration could provide ample ecological benefits and enhance environmental quality with minimal human maintenance and a small initial investment. Within Shanghai, in addition to the construction of green space in urban areas and greenbelts along the outer circle line, the near-natural forest method could be widely recommended for a number of purposes, including the following: restoring vegetation in garbage disposal fields, creating environmental protection forests in large industrial areas including those occupied by corporations such as the Shanghai Petrochemical Co., Ltd., planting coastal protection forests, creating forest networks within farms, and providing forests for ecological education and research. The method could be regarded as a new means of constructing urban forests, and as central to the construction of urban environments (Da and Song 2008). When planning urban river restoration projects, it is advisable to take advantage of near-natural restoration techniques adapted to local conditions. In most cases, the ecological functioning of river systems can be improved, while simultaneously minimizing the impacts of development and adding genuine social and economic value to the urban environment (Findlay and Taylor 2006).

However, there are still barriers to using near-natural methods in urban areas for restoration purposes, especially the need for management after the restoration, and unfavorable public perceptions. It is recommended that near-natural restoration be pursued through active collaboration with a range of other disciplines in order to improve restoration efforts on multiple fronts.



## References

Alexander M (1983) Introduction to soil microbiology. Translated by Agricultural Microbiology Research Group, Agricultural College, Guangxi University. Science Press, Beijing

Allen EB, Covington W, Falk DA (1997) Developing the conceptual basis for restoration ecology. Restor Ecol 5:275–276

Cao T, An L, Wang M, Lou YX, Yu YH, Wu JM, Zhu ZR, Qing YK, Glime J (2008) Spatial and temporal changes of heavy metal concentrations in mosses and its indication to the environments in the past 40 years in the city of Shanghai, China. Atmos Environ 42:5390–5402

Chen WM, Zhang QM, Dai SG (2009) The mutual allelopathy of *Vallisneria spiralis* Linn. and *Microcystis aeruginosa*. China Environ Sci 29(2):147–151

Ciesla WM, Donaubauer E (1994) Decline and dieback of trees and forests. A global overview. Food and agriculture organization of the United Nations (FAO) Forestry Paper 120. FAO, Rome

Da LJ, Song YC (2008) The construction of near-natural forests in the urban areas of Shanghai. In: Carreiro MM, Song YC; Wu J (eds) Springer Series on environmental management. Springer, New York, pp 420–432

Findlay SJ, Taylor MP (2006) Why rehabilitate urban river systems? Area 38(3):312–325

Gadgil PD, Bain J (1999) Vulnerability of planted forests to biotic and abiotic disturbances. New For 7:227–238

Gao JR (1999) Near natural control: torrent control engineering based on the landscape ecology. J Beijing For Univ 21(1):80–84 (in Chinese)

Gibson IAS, Jones T (1977) Monoculture as the origin of major forest pests and diseases. In: Cherret JM, Sagar GR (eds) Origins of pest, parasite, disease and weed problems. Blackwell, Oxford, pp 139–161

Graneli W, Solander D (1988) Influence of aquatic macrophytes on phosphorus cycling in lakes. Hydrobiologia 170:245–266

Hobbs RJ (1996) Towards a conceptual framework for restoration ecology. Restor Ecol 4:93–110

Huang JP, Wu CC (2002) Distribution of the microorganism groups in the air of forest area. Sci Silvae Sincae 38(2):173–176

Kan HD, Chen B, Chen CH, Fu Q, Chen M (2004) An evaluation of public health impact of ambient air pollution under various energy scenarios in Shanghai, China. Atmos Environ 38:95–102

Kloor K (2000) Restoring American's forests to their "natural" roots. Science 287:573–575

Kondolf GM (1995) Five elements for effective evaluation of stream restoration. Restor Ecol 3(2):133–136

Li L, Liu NN, Da LJ (2006) Nitrogen and phosphorus accumulation in *Iris tectorum* and *Acorus calamus* in eutrophic water. Environ Pollut Control 28(12):901–903 (in Chinese)

Li HJ, Ni LY, Cao T, Zhu LX (2008) Responses of *Vallisneria natans* to reduced light availability and nutrient enrichment. Acta Hydrobiol Sinica 32(2):225–230 (in Chinese)

Meuleman AFM, Beltman B, Scheffer RA (2004) Water pollution control by aquatic vegetation of treatment wetlands. Wetl Ecol Manag 12:459–471

Miyawaki A (1998) Restoration of urban green environments based on the theories of vegetation ecology. Ecol Eng 11:157–165

Miyawaki A (1999) Restoration of native forests by native trees. Plant Biotechno 16:15–25

Palmer MA, Ambrese RF, Poff NL (1997) Ecological theory and community restoration ecology. Restor Ecol 5:291–300

Ren WW, Zhong Y, Meligrana J, Anderson B, Watt WE, Chen JK, Leung HL (2003) Urbanization, land use, and water quality in Shanghai 1947–1996. Environ Int 29:649–659

Seidel K (1976) Macrophytes and water purification. In: Tourbier J, Pierson RW (eds) Biological control of water pollution. University of Pennsylvania Press, Philadelphia, pp 109–121

Seifert A (1938) Naturnaeherer Wasserbau. Deutsche Wasserwirtschaft 33(12):361–366

Shao HR, He QT (2000) Forest and Air Anion. World For Res 13(5):19–23 (in Chinese)

Starling F, Rocha AJA (1990) Experimental study of the impacts of planktivorous fishes on plankton community and eutrophication of a tropical Brazilian reservoir. Hydrobiologia 200: 581–591

Wang RQ, Fujiwara K, You H (2002) Theory and practices for forest vegetation restoration: native forest with native trees—introduction of the Madawaska's method for reconstruction of "environmental protection forest (ecological method to reforestation)". Acta Phytoecol Sinica 26(suppl):133–139 (in Chinese)

Wang JY, Da LJ, Song K, Li BL (2008) Temporal variations of surface water quality in urban, suburban and rural areas during rapid urbanization in Shanghai, China. Environ Pollut 152: 387–393

Wong MH, Bradshaw AD (2008) Progress in the reclamation of degraded land in China. In: Martin RP, Anthony JD (eds) Handbook of ecological restoration v. 2: Restoration in Practice. Cambridge University Press, Cambridge

Xu W, Wooster MJ, Grimmond CSB (2008) Modelling of urban sensible heat flux at multiple spatial scales: a demonstration using airborne hyperspectral imagery of Shanghai and a temperature-emissivity separation approach. Remote Sens Environ 112:3493–3510

Zhang Y, Zhao YW, Yang ZF, Chen B, Chen GQ (2009) Measurement and evaluation of the metabolic capacity of an urban ecosystem. Commun Nonlinear Sci Numer Simulat 14:1758–1765

Zhang KX, Wang R, Shen CC, Da LJ (2010) Temporal and spatial characteristics of the urban heat island during rapid urbanization in Shanghai, China. Environ Monit Assess 169:101–112

Zhu GP, Wang XR, Wang M, Qi S, Gao JR (2006) Advances on near natural comprehensive control of urban river. Sci Soil Water Conserv 4(1):92–97 (in Chinese)

# Chapter 18
# Sustainable Management of Urban Green Environments: Challenges and Opportunities

**Samuel Kiboi, Kazue Fujiwara, and Patrick Mutiso**

**Abstract** Urban green areas not only provide aesthetic qualities but also provide important ecosystem services in ever-shrinking habitats, and therefore need sustainable management practices. The western and northwestern parts of Nairobi are within an upland dry forest that stretched from Karura to Ngong forests with a characteristic vegetation composition. Much of that vegetation has been replaced by exotic species and, over time, the original indigenous tree species composition may be lost. No previous studies have profiled the local vegetation structure in Kenya and then used this knowledge to restore the urban green environment. We carried out studies in Karura and Ngong forests and used 16 carefully selected species to recreate a natural forest using the 'Miyawaki method' at the College of Biological and Physical Sciences of the University of Nairobi. In just 16 months the species have established extremely well, with the best performing species (*Ehretia cymosa*) growing to more than 210 cm from just about 43 cm. We expect to recreate a quasi-natural forest and use such studies and methods to restore urban green environments in Kenya.

**Keywords** Natural green environment restoration • Upland dry forests • Urban forest ecosystem services • Urban vegetation • Vegetation structure of Nairobi

---

S. Kiboi (✉) • P. Mutiso
School of Biological Sciences, University of Nairobi, Nairobi, Kenya
e-mail: samuel.kiboi@uonbi.ac.ke; mutiso_chalo@uonbi.ac.ke

K. Fujiwara
Graduate School in Nanobioscience, Yokohama City University, Yokohama, Japan
e-mail: kazue@ynu.ac.jp

N. Kaneko et al. (eds.), *Sustainable Living with Environmental Risks*,
DOI 10.1007/978-4-431-54804-1_18, 

## 18.1 Introduction

Urban green environments in developing countries especially are under constant pressures resulting from rapid urbanization, which can be sometimes unplanned. The greatest challenge today is to manage the environment in a sustainable way whilst offering pleasant surroundings for the urban dweller, and at the same time maintaining some natural areas. Nairobi, for example, has been ranked among the top ten cities in the world that have the biggest declines in liveability over 5 years with a score of −2.9 % according to the latest Global Liveability Survey of 140 cities worldwide (Economist Intelligence Unit 2013). The unit measured cost of living, health care, pollution, education, infrastructure, and green spaces to obtain the scores. In general, green environments, and specifically trees, offer an array of benefits that can be categorized broadly into: ecological benefits, architectural functions, climate moderation, and monetary benefits, as well as recreational and social values (TEEB 2011; Bolund and Hunhammar 1999).

## 18.2 Challenges in Achieving Sustainable Green Urban Environments

The challenges to sustainable green environments in urban areas are many and require a proactive approach and the cooperation of all citizenry. Apart from an effective regulatory system, an informed citizenry will keep the environment in better condition than those who have to be policed to adhere to standards. Some of the problems facing urban areas in Kenya are severe and common to many cities in the developing world, although they may vary in magnitude. Management of urban environments in the developed world is more effective, and hence the environments are cleaner and greener but may have experienced pollution challenges in the past, e.g., Yokohama in Japan had major pollution problems in the 1960s. Some of the major environmental challenges facing urban environments in Kenya are:

### *18.2.1 Waste Management*

Waste management is the biggest environmental problem in most urban areas and can lead to environmental pollution in many ways and affect drainage systems (Figs. 18.1, 18.2, and 18.3). The amount of solid waste generated per individual, homestead, or industry is usually high and requires a very comprehensive collection and disposal mechanism, ideally with facilities for sorting at source. In such cases organic waste is converted to manure or used to produce energy while most other types of waste are recycled. When it comes to liquid waste and effluents from industries, proper mechanisms are required to treat the waste to levels where no harm can result from discharging back to the environment.

**Fig. 18.1** An unregulated roadside garbage dump site in Nairobi where informal sorting of polythene is also undertaken

**Fig. 18.2** Industrial effluent discharge from a *blocked sewer line* showing waste treatment may be inadequate

**Fig. 18.3** Solid waste blocking drainage systems in Nairobi, Kenya

## *18.2.2 Air Pollution*

Air pollution comes from different sources in urban environments. The first major source is combustion of fossil fuels from vehicles and industries (Fig. 18.4). As the middle class continues to expand in developing countries, so do the number of vehicles on the roads. The maintenance of vehicles is usually a major contributor to how efficient their combustion process is and those that are badly maintained will emit exhaust gases that are not well combusted. This, combined with bad fuel quality, exacerbates the problem. Bad fuel quality may be due to use of outdated crude oil refining technology or due to illegal adulteration of fuel by mixing different types, e.g., diesel with kerosene.

Other sources of air pollution include burning of waste and, perhaps worst of all, rubber (e.g., burning old tires to remove the steel ply inside) and plastic waste, especially in unmanaged dumpsites. Rubber and plastics release substantial amounts of toxic gases into the air which not only contributes to global warming but also to health problems among those inhaling the air. Open tire fire emissions include pollutants, such as particulates, carbon monoxide (CO), sulfur oxides ($SO_2$), oxides of nitrogen ($NO_x$), and volatile organic compounds (VOCs). They also include hazardous air pollutants (HAPs), such as polynuclear aromatic hydrocarbons (PAHs), dioxins, furans, hydrogen chloride, benzene, polychlorinated biphenyls (PCBs), and metals such as arsenic, cadmium, nickel, zinc, mercury, chromium, and vanadium (Lemieuxa et al. 2004).

**Fig. 18.4** Emission of gases from fossil fuels contribute to the amount and concentration of greenhouse gases in cities such as Nairobi

Lack of grass cover or properly paved roads and walkways leaves the soil exposed and constant traction by vehicles and human traffic results in dusty conditions that lead to respiratory diseases, as well as dusty buildings and installations. Urban dust is more likely to carry harmful substances including microbes, because often drainage systems block due to runoff soil and organic waste flooding the roads, and the micro-organisms can then be spread via dust. Therefore, if well-managed green areas and paved surfaces exist, these problems would be minimized.

### *18.2.3 Infrastructure*

Infrastructure development is a key measure of urban growth. All developments should ideally be well planned with the necessary regulatory approvals including minimizing environmental impacts by undergoing an impact evaluation. Measures to mitigate environmental impact should be put in place during project implementation and maintained after commissioning. In Kenya for example, the National Environmental Management Authority (NEMA) or Environmental and Social Impact Assessment (ESIA) - (depending on scope) has to approve all projects after an Environmental Impact Assessment (EIA) has been conducted. Despite the regulations being followed, however, compliance challenges exist, especially in

**Fig. 18.5** Proper designation of clear zones along roadsides can help avoid such vegetation disturbance and better management for outdoor adversing space

ensuring that what is approved on paper is actually implemented during and after inception of the project. The capacity of such institutions may be limited such that each project is not monitored and assessed as it ideally should be.

An example of an urban environmental problem affecting green due to infrastructure development is the information technology industry. In the last 5 years or so, Kenya and the region as a whole have invested heavily in fiber optic connectivity with up to three international cable landings connecting Kenya and the rest of the world. Most if not all major towns now have a connection using the fiber optic network. Companies that have invested in the cable are now racing to connect as many buildings as possible and the negative effect of this is that there is no common cabling approach. This means that different companies dig up the same routes at different times and the result is repeated disturbance, while environmental restoration is not always to the previous standard or better.

Another example of challenges to green urban areas is the billboard advertising industry. While strategic positioning is important in brand visibility, urban greening programs could affect visibility over time as the trees mature. Observations around Nairobi show that this is countered by trimming trees, which is not ideal for coexistence between outdoor advertising industry and sound ecologically friendly urban environmental management (Fig. 18.5). This is a severe problem that needs to be addressed by the city management authorities and the advertising companies in conjunction with the NEMA.

## 18.3 Opportunities for Sustainable Green Urban Areas

Trees offer opportunities to create multi-strata natural areas and absorb 25–30 times more $CO_2$ than mono-strata grass surface areas, and therefore a scientifically guided restoration or reconstruction of green environments such as vegetation ecology or ecotechnology is desirable (Miyawaki 1998). Ecotechnology is a technological means of ecosystem management based on deep ecological understanding, to minimize costs of management measures and their harm to the environment (Straskraba 1993). Ecotechnology uses a vegetation ecology approach to build multi-strata greenery (e.g., canopy, understory, shrub, and herb layers) to fulfill the ecological potential of the area. Forests have different ecological functions and urban greening programs should not rely on the functions of green vegetation in general, but should address the suitability of different plants in light of the complexities of urban design and management. Such complexities may include: (1) open spaces such as parks, (2) areas along roads and walkways, (3) tracts of land along riverine areas and in wetlands, (4) mixed-use areas such as those around and within residential districts and other built-up areas, and (5) office and industrial zones.

Trees are the most memorable aspect of a roadside planting design. They have an appropriate scale for a road corridor, are clearly noticed when travelling and are the best means for ameliorating the hard built elements of the road corridor. Subject to their safe use, they should be the primary element of a landscape design. Trees should, however, be used selectively in a corridor. For example they should not obscure expansive views and they should be located carefully and deliberately, outside clear zones and away from utilities (Chang and Collins 2008)

Each urban district occurs in a distinct ecological area with unique naturally occurring plant species. These are species that have long adapted in these ecosystems and established their own niches in the communities that they occur in. In selecting tree species to plant in urban areas, this should be a key consideration to maintain natural environments. In many upcoming cities in the developing world, and in Kenya in particular, there seem to be no scientifically informed criteria for urban greening programs, and most of the time exotic species are planted. This may mainly be due to their visual appeal, availability, and growth under a wide breadth of ecological conditions (e.g., *Grevilia robusta*). Indigenous species are usually overlooked and this could be mainly because of insufficient understanding of their local suitability, optimum combinations, and interaction with other species, as well as impact on the infrastructure, among other considerations. Nonetheless, some other countries have made significant progress in urban green environments based on theories of vegetation ecology (Miyawaki 1998; Muller and Fujiwara 1998).

Some key benefits of green urban environments include carbon assimilation, disaster prevention and reduction, beautification and nutrition.

### *18.3.1 Carbon Assimilation*

Urban environments are major sources of carbon dioxide ($CO_2$) and other green house gases (methane, nitrous oxide, and fluorinated gases) that contribute to global warming, a major contributor to climate change today. $CO_2$ is the primary greenhouse gas emitted through human activities. The majority of GHGs emanate from burning fossil fuels, mainly from motor vehicles and industry (International Energy Agency 2012). These emissions have risen steeply over the last century with some cities now being covered in smog that not only reduces visibility, but also reduces air quality, generating air pollution which in turn can cause major respiratory issues and associated health complications. Smog generally can affect plant development and human health, as well as cause damage to materials such as rubber, textiles, and paint (Marcella et al. 1957; Rani et al. 2011). Three major outdoor air pollution problems are industrial smog from burning coal, photochemical smog from motor vehicle and industrial emissions, and acid deposition from coal burning and motor vehicle exhaust (Rani et al. 2011). A wide range of experts have advocated decreasing individual carbon footprints and investing billions to reduce the risks of a major change in the earth's environment (Stern 2008). Green environments can help absorb the $CO_2$ in the air which can be converted into stored carbon through the process of photosynthesis. Different plant species vary in the amount of $CO_2$ they can absorb and this variation is determined by various morphological and physiological characteristics of the plants as well as land use (Houghton 1989). Some characteristics that can help determine which plants would have high carbon assimilation rates include growth rate, leaf area, wood density, and seasonal vegetation changes, including whether they are deciduous or not.

### *18.3.2 Disaster Prevention and Reduction*

Disaster prevention and reduction is a key contribution of green areas in urban environments when carefully planned. Experience from Japan has shown that areas with trees along elevated highways and railways have proved very important as a disaster prevention and reduction method because the trees form important barriers to fires and offer support from total collapse due to earthquakes or tsunamis (Miyawaki 1998). In Kenya, congestion of highways in urban areas such as Nairobi is forcing planners to consider elevated highways as a way of increasing capacity to handle vehicle traffic efficiently. Therefore, an integrated transport management strategy should incorporate planting the right species that can mitigate such incidents if they were ever to occur.

Headlight glare from opposing traffic can cause potential safety problems and plants can serve to reduce the glare during night time driving. The most favored design of highways today is to have dual carriageways that can handle several lanes of traffic going each way. Even when the carriageways are separated by ample distance between them you will find that headlamp glare from motor vehicles will always affect other motorists around corners or bends. Glare can be reduced by the use of wide medians, separate alignments, earth mounds, plants, concrete barriers, and glare screens (WSDOT Design Manual 2013). Long-term maintenance should be considered when selecting the treatment for glare but some solutions can be expensive (e.g., glare screens). Plants can be used to create natural light barriers between the highways and consequently block light from headlamps, making night time driving safer and more pleasant. However, such greening programs need to be based on informed decisions regarding what species to plant where, and at what distances from the roads and road junctions. For example, if wrongly planted or mixed, trees can block the driver's view at junctions and clear zones, thereby increasing risk of accidents, or of impact if the driver loses control of the motor vehicle. It is essential to understand the branching system and/or strength to ensure selected species do not break off easily, which is important in withstanding high winds, and consequently enhancing the safety of pedestrians, motorists, and utility lines. Some tree species, e.g., many species of Eucalyptus will snap during the rainy season when they cannot support the large volume of water they take up during this time combined with the effects of higher wind speeds or wind gushes.

The other importance of having vegetation in between dual carriageways or along highways is that it can act as a barrier when accidents occur. While tree removal may be beneficial to reduce the impacts of driving errors (e.g., angle crashes), appropriate vegetation may help to reduce speed and magnitude of impact in case of an accident. This can be achieved through frangible planting—planting which breaks under the impact of a motor vehicle (and hence helps to stop the vehicle). Generally trees and shrubs with a mature trunk diameter of less than 100 mm at around 500 mm above ground level are considered frangible. Vegetation can act as softer barriers than concrete or metal and therefore reduce impacts during an accident and increase the chance that injuries or loss of life are minimized.

### 18.3.3 Beautification

Since time immemorial, man has used different types of plants for beautification ranging from herbs and shrubs to trees. Plants of different species produce a wide variety of flowers with pleasant odors and colors. Flowers, being the evolutionary adaptation to help plants in pollination therefore have different shapes, colors, scents, and rewards such as pollen or nectar to attract specialized animal pollinators. Man has proceeded to breed and domesticate other flowers whose only purpose is beauty, usually deriving from the flower and/or leaf color. These plants are used both indoors and outdoors, and grown in a variety of ways including in pots, on walls and rooftops, in lawns and hedges, and in home gardens, as well as in parks and urban forests. Local vegetation that is well understood (e.g., in terms of flowering pattern and cycle, scents and vegetative growth) can be selected and incorporated in urban greening programs and serve the additional benefit of beautification.

### *18.3.4 Nutrition (Fruits)*

Urban vegetation planning can include local and exotic fruit species. This can be in home gardens, public parks, and within compounds of institutions and office complexes. In Sweden for example, it is not uncommon to find fruit trees such as apples in public parks and people are free to pluck and consume as they relax or pass through these areas. In Kenya, fruits such as mangoes, avocadoes, plums, and coconuts can be incorporated in urban greening programs. At present, it is popular to grow fruit trees in urban home gardens but not common in public places.

## 18.4 Case Study: Opportunities in Urban Environments

### *18.4.1 Restoration of an Urban Green Environment Using Potential Natural Vegetation at the University of Nairobi, Kenya*

The College of Biological and Physical Sciences of the University of Nairobi sits between Kirichwa Kubwa and Kangemi Rivers, and can be described as the zone where altitudinal cline starts accompanied by obvious remnants of upland dry forest vegetation to the west, south west, and north west, and separates the savannah landscape to the east and south east. This includes the riverine tree species such as Syzygium and Albizia. It is a unique and picturesque location that Ewart Scott Grogan (1874–1967), one of the pioneer colonial settlers, chose to build a home. Remnants of original vegetation show that the upland dry forest was continuous and spread from Karura upwards towards Ngong forests. However, only a few of the original species remain since the colonial settlers planted exotic species such as Eucalyptus (to drain the wetlands next to the two rivers), pine, and other species such as Jacaranda for beautification purposes. In addition, substantial infrastructure developments have taken place with the former residence converted into a campus of the University of Nairobi. The college is one of the few remaining areas of Nairobi with a natural feel and enjoys a clean and fresh environment compared to the city center barely 2 km away.

It therefore offers an opportunity not only to preserve the environment but also to restore it with original natural vegetation that has evolved in place for millions of years, while still offering a pristine educational environment for present and future generations.

To restore an environment with potential natural vegetation, one needs to understand the kind of vegetation that existed before human disturbance. Since 2007, our team led by Professor Kazue Fujiwara has conducted some intensive studies in Karura and Ngong forests with the aim of understanding vegetation composition and the social relationships among species. From these studies, we were able to identify common species that form all levels of vegetation in a forest from the canopy (T1) and understory (T2) canopies down to shrub and herb layers, and use this mix to create a natural forest by planting seedlings using the Miyawaki method (Miyawaki 1998). The selected species were *Shrebera alata*, *Rawsonia lucida*, *Cassipourea malosana*, *Vepris simplicifolia*, *Drypetes gerrardii*, *Elaeodendron buchananii*, *Croton megalocarpus*, *Brachylaena huillensis*, *Calodendrum capense*, *Ficus thonningii*, *Warburgia ugandensis*, *Olea europaea* ssp. *Africana*, *Olea capensis* ssp. *Hochestetteri*, *Ehretia cymosa*, *Markhamea lutea* and *Cordia africana*. The evenness of the above species was matched to ratios close to their natural distributions as we had found them in the natural forest studies such that, for example, there were more *W. ugandensis*, *C. capense*, *C. megalocarpus*, *B. huillensis*, and *C. africana* in the mix than *S. alata*, *E. cymosa*, *F. thonningii,* and *R. lucida*.

In designing an urban forest, understanding the species to plant is critical but also other important considerations come into play. Urban areas have zones such as recreational areas, roads, buildings, rivers, etc., each with their own characteristics, and therefore specific considerations and informed decisions should be applied in each case. For example, planning of tree planting in recreational areas such as parks needs to take into consideration spatial arrangements to enhance movements of people, as well as placement of benches. Therefore the pattern and intensity of planting must be well planned. In addition, the species selected should not break off easily or be uprooted by strong winds because of the potential dangers of them falling on people. In the case of Chiromo campus, we considered that the area chosen was also used for recreational purposes and so we could not plant a continuous forest. We decided to take advantage of the less utilized part of the area which formed step-like contours along the gradient of the slope (Fig. 18.6). Trees also do better on a slope than on a flat area, mainly because slopes have better drainage.

The site was prepared by clearing grass and bushes, and then holes of at least 0.3 m wide by 0.6 m deep were dug at a density of 3 holes/m$^2$. Seedlings were sourced from the local community groups as well as the Kenya Forestry Services nurseries. The 16 species were then mixed randomly and placed in each hole, with over 2,000 seedlings planted. The planting was undertaken by the university community, including members of staff and students, along with Japanese researchers and volunteers (Professor Miyawaki and Professor Fujiwara, as well as other scientists and volunteers). The student participation was very important because they were able to learn firsthand, as well as being able to take the experience out of the University when they finished their degrees. On conclusion of the planting, mulching from cut grass was placed around each seedling to prevent water loss and soil erosion. Management (weeding and replacement of any dead seedlings) has been ongoing since then and is expected to continue for a total of 3 years, at which point the seedlings are expected to be big enough.

### 18.4.2 Seedling Performance

The height of the seedlings was measured in September 2013 after 16 months of growth and the performance had been very good (Fig. 18.7). Some of the species had grown very fast with *S. alata* attaining the best mean growth of 211 cm (followed

**Fig. 18.6** Newly planted site at the University of Nairobi with seedlings about 43 cm in height after 2 months (Photo dated: 28.06.2012)

**Fig. 18.7** Same site after 16 months with some seedlings exceeding 2 m in height (*Inset*: Measuring height of *Cordia africana*-almost twice as tall as the person holding the measuring ruler)

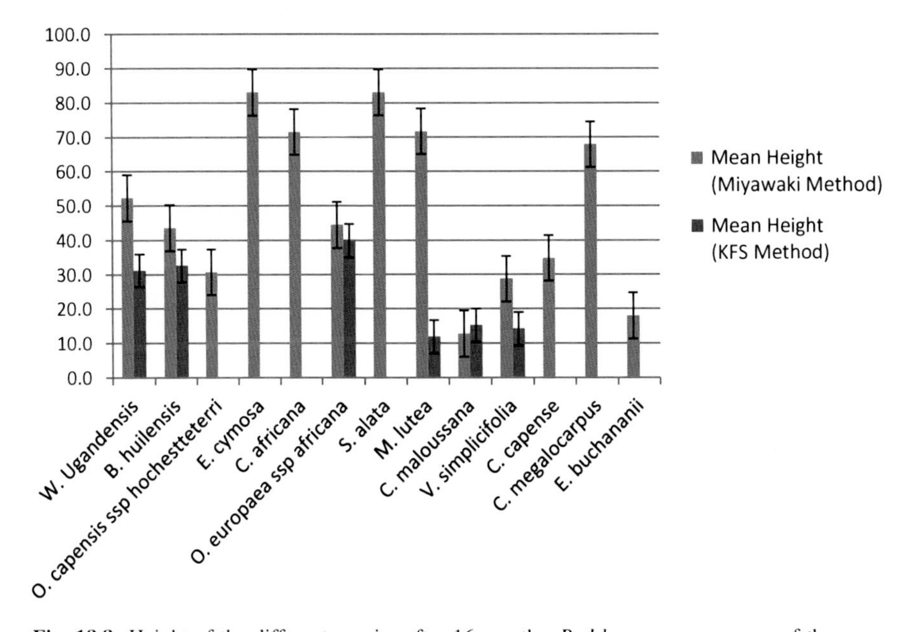


**Fig. 18.8** Height of the different species after 16 months. *Red bars* compare some of the same species planted at the same time at a low density using the regular method

closely by *E. cymosa* (210 cm), *M. lutea* (182 cm) and *C. Africana* (181 cm)) (Fig. 18.8). The slowest mean growth was that of *C. malousana* (32.5 cm) and *E. buchananii* (45.7 cm). The rest of the species had median growths varying between 71 and 132 cm. Compared to the traditional planting method with a spacing of 1 $m^2$ between the seedlings, the Miyawaki method showed better performance in the few species that could be compared (Fig. 18.8).

## 18.5 Conclusion

Scientifically informed decisions can help restore urban green vegetation by helping in identification of the right species and their combinations. The method of rehabilitation is also important as it is evident here that the Miyawaki method favors faster establishment of vegetation and therefore it is possible to regenerate a quasi natural forest over much shorter temporal scales. Involving communities, the public, or students is important in long term success since they learn and take the experiences to different parts of a country as well as taking ownership and practicing sustainable environmental practices on their own initiative.

## References


Bolund P, Hunhammar S (1999) Ecosystem services in urban areas. Ecol Econ 29:293–301

Chang J, Collins G (2008) Landscape guideline. Landscape design and maintenance guidelines to improve the quality, safety and cost effectiveness of road corridor planting and seeding. RTA, Sydney

Houghton RA (1989) The long-term flux of carbon to the atmosphere from changes in land use. Extended abstract from the third international conference on analysis and evaluation of atmospheric $CO_2$ data present and past. Hinterzarten, 16–20 October 1989. Environmental Pollution, Monitoring and Research Programme 59. World Meteorological Organization

International Energy Agency (2012) $CO_2$ Emissions from fuel combustion highlights. IEA Publications, Paris

Lemieuxa PM, Lutesb CC, Santoiann DA (2004) Emissions of organic air toxics from open burning: a comprehensive review. Prog Energ Combust 30:1–32

Marcella J, Noble W, Went FW (1957) The standardization of *Poa Annua* as an indicator of smog concentrations. I. Effects of temperature, photoperiod, and light intensity during growth of the test-plants. Plant Physiol 32:576–586

Miyawaki A (1998) Restoration of urban green environments based on the theories of vegetation ecology. Ecol Eng 11:157–165

Muller N, Fujiwara K (1998) Biotope mapping and Nature conservation in cities Part 2: Results of a pilot study in the urban agglomeration of Tokyo (Yokohama City). Bull Inst Environ Sci Technol Yokohama Natn Univ 24:97–119

Rani B, Singh U, Chuhan AK, Sharma D, Maheshwari R (2011) Photochemical smog pollution and its mitigation measures. J Adv Sci Res 2:28–33

Stern N (2008) The economics of climate change. Am Econ Rev 98(2):1–37

Straskraba M (1993) Ecotechnology as a new means for environmental management. Ecol Eng 2:311–331

The Economics of Ecosystems and Biodiversity (TEEB) (2011) TEEB manual for cities: ecosystem services in urban management. www.teebweb.org

The Economist Intelligence Unit (2013) A summary of the liveability ranking and overview. The Economist Group, London, UK

WSDOT Design Manual (2013) Roadside safety. 1600-01-1600-22

# Part IV
# Policy, Institutuinal and Capacity Development, Education and IME System

# Chapter 19
# Participatory Sustainability Research for Risk Management and Leadership Development

**Masanori Kobayashi**

**Abstract** Effective risk management and sustainability promotion require proper assessment of the environment and social capacity for managing the environment. National governments and international agencies provide monitoring data on the environment such as data relating to air and water quality, forest cover, land, biodiversity, and waste management. While local communities and stakeholders need to play a vital role in managing risks and promoting sustainability at the field level, however, they often lack scientific information. Instead they rely on the experiential and observation-based information that is often the most useful in communicating to other community members and stakeholders. Participatory assessment can therefore provide a useful tool for community members and stakeholders to comprehend environmental risks and challenges in promoting sustainability. Meanwhile, the feedback from the communities and stakeholders constitutes useful information for decision-makers and practitioners to plan and facilitate transformation in policies and institutions in order to improve environmental risk management and promote sustainability.

**Keywords** Environmental risks • Leadership development • Social capacity assessment • Stakeholder participation • Sustainability

M. Kobayashi (✉)
Graduate School of Environment and Information Sciences,
Yokohama National University, Yokohama, Japan
e-mail: m-kobayashi@ynu.ac.jp

N. Kaneko et al. (eds.), *Sustainable Living with Environmental Risks*,
DOI 10.1007/978-4-431-54804-1_19, 

## 19.1 Introduction: Environment and Social Capacity Assessment for Effective Environmental Risk and Sustainable Resource Management

It has long been acknowledged that local communities and stakeholders have a vital role to play in protecting the environment, managing risks, and promoting sustainable use of natural resources, and environmental issues are best handled with the participation of all citizens concerned, as reaffirmed in the Rio Declaration adopted at the United Nations Conference on Environment and Development (UNGA 1992). Ostrom (1990) refers to participatory environmental management by characterizing it as management of common pool resources by non-governmental actors. Ostrom (2000) further articulates the features of self-governance of natural resources. Ostrom asserts that the presence of leaders or entrepreneurs is an important factor in instigating social mobilization, structuring governance mechanisms, and promoting collective actions to manage the environment and natural resources.

When developing and operationalizing self-governing mechanisms for environmental management or sustainable natural resource use, the role to be played by leaders or entrepreneurs is paramount, but emphasis is also given to the utility of involving external facilitators in similar processes (APFED 2010). The latter supplement the leaders or entrepreneurs, who can be endogenous. External facilitators are involved in raising awareness among local community members and stakeholders, organizing them, and institutionalizing collaborative management of the environment and natural resources. They can play an instrumental role in identifying and providing options for interventions in addressing environmental risks or sustainability challenges. It is however important to note the caveat that the external facilitators also need to understand the local conditions and context (Adandedjan and Niang 2006). They are not supposed to impose their own preconceived notions of collective actions on community members and stakeholders (Sow and Adjibade 2006). It must be the local community members and stakeholders who make the final decisions on the modalities of collective actions.

It is a core purpose of sustainability science to understand the mechanisms of natural resource use and impacts on the environment and livelihoods. To carry out effective sustainability science, it has been suggested that multi-disciplinary expertise should be brought together to facilitate in-depth transformation of the way scientific research is organized (Dedeurwaerdere 2013). Integration of science and knowledge about natural and social systems has evolved to become sustainability science (Rockström et al. 2009; Blackstock et al. 2007; Yasunari 2013) and a platform has been established to promote science-policy interface on sustainability (Takeuchi 2013). Sustainability science is said to consist of two key components, namely a descriptive analytical mode based on an advanced form of complex system analysis, and a transformation mode oriented toward developing practical solutions for sustainability problems. An increasing emphasis is now given to the latter component to facilitate a socio-economic transition toward achieving stronger sustainability (Dedeurwaerdere 2013).

If the participatory appraisal for the environment and ecosystems can be undertaken using a simple and indicative assessment approach, such a process can provide a useful tool as well as a step toward commencing dialogues for stakeholder/community leaders, scientists, and practitioners. Such dialogues can address how to improve environment/ecosystem management and livelihoods as a preliminary stage in participatory learning and action (PLA), or participatory sustainability science (PSS) research. The participatory appraisal process can also provide a meaningful opportunity for students to acquire holistic viewpoints on sustainability and develop the analytical and facilitation skills required for environmental leaders by confronting reality and being required to comprehend the complexity and dynamism intrinsic to sustainability conundrums. In practice, however, PLA or PSS research does not always lead to the intended outcome due to either poor conflict resolution or insufficient facilitation (Blackstock et al. 2007) and there are a number of challenges in integrating PLA or PSS into environmental leadership training programs. This paper aims to outline the genesis and evolution of PLA or PSS, its advantages, and the challenges in applying it in environmental leadership development at universities.

## 19.2 Genesis and Evolution of Participatory Sustainability Science

There has been a looming question about the extent to which researchers and university scholars can carry out sustainability science with greater impacts on society. The participatory rural appraisal (PRA) approach that spread and evolved in the 1980s developed as a method to learn about rural livelihoods from, with, and by rural people (Chambers 1994a) or to enable local people to share, enhance, and analyze their knowledge of life and environmental conditions, to plan, and to act (Chambers 1994b). PRA emanated from preceding currents such as activist participatory research that aims to enhance people's awareness and confidence and to empower their actions. With the increased proactive involvement of local people in learning processes, the scope of PRA was expanded and it was renamed "participatory learning and action" in the 1990s (Chambers 2007). In addition, a number of other developments have emerged and been promoted as PSS. These include generation of knowledge about sustainability from multiple stakeholders, attempts to prompt changes in personal and institutional behaviors, and a transition toward sustainability (Blackstock et al. 2007). The evaluation framework for PSS research is structured with multiple components including the level and modalities of stakeholder collaboration that changed people's behaviors, norms, and culture (Blackstock et al. 2007).

Lately, a participatory sustainability research framework with an articulate focus on the environment, natural resources, and local livelihoods has been applied. A variety of multi-ecosystem service assessment techniques have been developed and practiced (Kelvin et al. 2013). Multi-ecosystem service assessment techniques are mainly based on the framework provided by Millennium Ecosystem Assessment 2005.

**Table 19.1** Ecosystem services and drivers of change

| Ecosystem services | Direct drivers | Indirect drivers |
|---|---|---|
| 1. Provisioning (e.g., food, water, fiber and fuel),<br>2. Regulating (e.g., climate regulation, water, and disease),<br>3. Cultural (e.g., spiritual, aesthetic, recreational and educational) and<br>4. Supporting (e.g., primary production and soil formation) | 1. Changes in local land use and cover,<br>2. Species introduction or removal,<br>3. Technology adaptation and use,<br>4. External inputs (e.g., fertilizer use, pest control, and irrigation),<br>5. Harvest and resource consumption,<br>6. Climate change and<br>7. Natural, physical and biological drivers (e.g., evolution, volcanoes) | 1. Demographic,<br>2. Economic (e.g., globalization, trade, market, and policy framework),<br>3. Socio-political (e.g., governance, institutional and legal framework),<br>4. Science and technology and<br>5. Cultural and religious (e.g., beliefs, consumption choices) |

Millennium Ecosystem Assessment (2005)

Millennium Ecosystem Assessment separates the various ecosystem services into four categories, namely (1) provisioning, (2) regulating, (3) cultural, and (4) supporting (Table 19.1). Millennium Ecosystem Assessment presents the direct and indirect drivers that affect ecosystems and delineates the mutual interactions—positive and negative—between ecosystems and humans on different spatial and temporal scales.

Millennium Ecosystem Assessment assessed the changes in major ecosystem service components as "enhanced," "degraded," or "no change in net." The drivers' impacts were assessed as "low," "moderate," "high," or "very high." Meanwhile, the drivers were assessed as "decreasing," "continuing," "increasing," or "rapidly increasing" their impacts. This approach of indicating the drivers' impacts and their changes in simplified terms is deemed pragmatic, particularly in participatory environment/ecosystem assessment. In addition, the Millennium Ecosystem Assessment approach involved assessing the response options for managing ecosystems with particular attention to "government," "business," and "civil society."

Matsuoka et al. (2008) scrutinize the capacity of these three social actors using "social capacity assessment" and "actor-factor analysis." The government, business, and civil society are assessed in terms of their capacity to fulfill the required social system functions with an emphasis particularly on (1) policies and measures, (2) human and organizational resources, and (3) knowledge and technology.

For the United Nations Conference on Sustainable Development (Rio+20 Summit) held in Rio de Janeiro, Brazil, in June 2012, the institutional framework for sustainable development was chosen as one of the two main themes, together with green economy (UNGA 2012). In the outcome document adopted at the Rio+20 Summit, an unequivocal statement was included in paragraph 99 calling for actions to promote access to information, public participation, and justice in environmental matters. While the importance of participation is already emphasized and a number of initiatives have been launched to promote access to information, participation in decision-making, and judicial proceedings over environmental matters, many countries still lag behind in developing the required legislation and enforcement mechanisms (Kobayashi 2012). In 2002, the World Resource Institute undertook the first assessment of its kind on access to information, participation in decision making, and judicial proceedings over environmental matters (Petkova et al. 2002). Governments in Asia were urged anew to make environmental information available more proactively, in a format comprehensible to the public, and without the need for a request. At the same time they were urged to enhance transparency through information disclosure to enable the public to participate in decision-making (WRI 2013).

The prioritization and selection of factors that are important in environment and social capacity assessment vary depending on the local conditions and the context of the intended participatory sustainability research (Blackstock et al. 2007). Collective self-reflection through interaction and dialogue among diverse participants supports a cyclical process of observation, analysis, planning, implementation, monitoring, and reviewing based on experiential learning and fosters adaptive management of the environment and ecosystems (Mackenzie et al. 2012). The following section will present further analysis regarding the applicability and challenges of participatory sustainability research. The analysis is based on the experience of implementing a joint research program undertaken by Yokohama National University and the University of Antananarivo.

## 19.3 Achievements and Challenges in Applying Participatory Sustainability Research: Field Study in Madagascar

As a part of the Leadership Development Programme for Sustainable Living with Environmental Risks (the SLER programme), Yokohama National University undertook joint field studies with the University of Antananarivo during 2011–2013. The second joint field study was undertaken from October 27 to November 10, 2012. Preparatory consultation meetings and an outcome presentation symposium took place in Antananarivo on October 28–30 and November 7, respectively, with the field visits from October 30 to November 6. A total of 17 participants, including seven faculty members and ten graduate students from two universities, participated in the field visits (YNU-SLER 2013a, b). The main objective of the field visits was to understand environmental risks and their mechanisms and to observe the intervention measures for reducing risks and promoting sustainability in Madagascar. The participants were separated into two groups and Group A visited the mid-east of Madagascar including Andasibe, Ambatondrazaka, and Alaotra Lake. Meanwhile, Group B visited the north-east of Madagascar including Andapa, Sambava, and Antalaha (the so-called SAVA region—named after the initials of its main cities, namely Sambava, Antalaha, Vohémar, and Andapa ).

The groups undertook (1) an environment/ecosystem assessment, (2) a soil survey, and (3) a social/stakeholder survey (interviews and questionnaire). However, the purpose of this paper is not to present the details of their field research findings. Instead, it

**Table 19.2** Ecosystem assessment in Madagascar—Highlights

| Ecosystem | Ecosystem service (resource) | Assessment | Future condition of resources | Reason for the condition |
|---|---|---|---|---|
| Forest | Fruits | ++ | ↘ | Increasing tree cutting |
| | Fuel | +++ | ↘ | Expanding forest fires |
| | Building material | ++ | → | |
| | Craft products | n/a | → | |
| | Honey | + | ↘ | |
| | Medicine | + | → | |
| | Tourism | n/a | → | |
| Coast | Seafood | +++ | ↘ | Sedimentation |
| | Tourism | n/a | → | Deforestation<br>Increasing fisherman |
| Crop land | Crop | +++ | ↘ | Land erosion<br>Low input |

+++: Abundant; ++: Adequate; +: occasionally available; –: poor/rare; n/a data not available; ↑ Rapidly increasing; ↗ Increasing; → No change; ↘ Decreasing; ↓ Rapidly decreasing
Developed from Miura, 2013, Assessment of natural resource and social capacity in Madagascar, unpublished

aims to highlight the potential and challenges in undertaking participatory sustainability research as a part of the environmental leadership development program.

Both groups had already gained an overall understanding of the trends in terms of environmental and ecosystem degradation through pre-visit literature review and interviews with stakeholders. Deforestation for charcoal production is prevalent, as is land reclamation to expand paddy and crop fields, and lumber extraction. In Ambatondrazaka, it is evident that deforestation causes siltation and sedimentation in irrigation channels and paddy fields, and reduces paddy field productivity. Due to the reduced rice productivity coupled with an increasing demand for rice, local farmers and villagers resort to forest destruction and land reclamation. A series of such behaviors constitutes a so-called "poverty-environment degradation vicious cycle" (Aggrecy et al. 2010). Based on the outcomes of the field surveys, assessments were also made of overall ecosystem service status and trends covering major ecosystem services, although the assessments were not exhaustive (Table 19.2).

Soil surveys were undertaken to examine the potential correlation between the land use practice and soil conditions, with a particular emphasis on whether land was excavated or under non-tillage cultivation (YNU-SLER 2013b). The groups conducted their soil survey on the assumption that soil conditions would be improved by conservational land use in the form of non-tillage farming or grass coverage of slopes for soil erosion control. It was assumed that the positive impacts on soil conditions could be shown by: (1) an optimal level of soil pH, or marginal deviation from the optimal pH level (expressed by "6.5—pH" when pH was lower than 6.5, and by "0" when pH was in a range of 6.5–7); (2) high electric conductivity; (3) high transparency of water containing dissolved soil; (4) a low level of soil hardness; and (5) low soil weight per unit. The results obtained in the soil inspection in the SAVA region concurred with the original assumption, as shown in Table 19.3.

**Table 19.3** Soil analysis—SAVA region Correlation with non-tillage land use history—24 locations of 12 sites

| Variables | Data | Expectation | Correlation coefficient |
|---|---|---|---|
| pH | 6.5-pH (pH<6.5)<br>0 when pH is 6.5–7 | Low | 0.2795 |
| Electric conductivity | mS/cm (millisiemens per centimeter) | High | 0.3069 |
| Transparency of dissolving water | Measures from 1 to 9 with 9 for highest transparency | High | 0.3941 |
| Hardness | Higher number for higher hardness | Low | 0.2677 |
| Weight/100 cc | Gram | Low | 0.2726 |

Optimal range of pH is between 6.5 and 7 (Thermo Fisher Scientific n.d.)
*Note*: pH was less than 7 in all sites

**Table 19.4** Social survey—Ambatondrazaka and SAVA areas Highlights (1)

| Average monthly salary (USD 1 = MGA 2,200) | <20 k | 20–60 k | 60–200 k | 200–400 k | 400–800 k | >800 k |
|---|---|---|---|---|---|---|
| Ambatondrazaka area (n = 32) | | | | | | |
| Household no. | 7 | 12 | 9 | 3 | 1 | 0 |
| Accumulative % (n = 32) | 22 | 59 | 88 | 97 | 100 | NA |
| SAVA area (n = 32) | | | | | | |
| Household no. | 9 | 12 | 6 | 3 | 0 | 0 |
| Accumulative % (n = 32) | 30 | 70 | 90 | 100 | NA | NA |
| **Number of children** | ≦2 | 3–4 | 5–6 | 7–8 | 9 | 10 |
| Ambatondrazaka area (n = 32) | | | | | | |
| Household no. | 15 | 10 | 7 | 3 | NA | NA |
| Average 3.2/household | | | | | | |
| SAVA area (n = 32) | | | | | | |
| Household No. | 7 | 8 | 10 | 3 | 2 | 1 |
| Average 4.6/household | | | | | | |
| **Education** | None | Elementary | Elementary + Junior High | High school | Voca-tional | Univer-sity |
| Ambatondrazaka area (n = 35) | 2 | 7 | 6 | 11 | 7 | 2 |
| SAVA area (n = 27) | 1 | 9 | 11 | 6 | 0 | 0 |

The soil samples taken at the site with the history of non-tillage land use demonstrated the projected coefficiency with the soil condition factors.

The groups also undertook questionnaire surveys to understand the socio-economic conditions of people living in the sites visited. The questions encompassed household income level, the number of children in each household, and education history. The questions also included a self-assessment of the environment and invited suggestions for possible interventions to arrest environmental degradation and promote the restoration of the environment and ecosystems (Tables 19.4 and 19.5).

**Table 19.5** Social survey—Ambatondrazaka and SAVA areas Highlights (2)

| | Ambatondrazaka area | | | SAVA area | | |
|---|---|---|---|---|---|---|
| | Agree | Disagree | Do not know | Agree | Disagree | Do not know |
| Promoting environmental education | 33 | 1 | 0 | 31 | 0 | 0 |
| Enforcing penalty on illegal logging | 31 | 0 | 0 | 31 | 1 | 0 |
| Giving more budget for government's monitoring of the environment | 22 | 7 | 1 | 28 | 1 | 1 |
| Raising tax on the sales of charcoal | 13 | 17 | 0 | 11 | 6 | 3 |
| Increasing the tax on land use | 4 | 23 | 0 | 6 | 16 | 5 |
| Increasing the government's subsidy for tree plantation | 30 | 1 | 0 | 31 | 0 | 0 |
| Encouraging private sector support for tree plantation | 29 | 0 | 2 | 23 | 2 | 0 |
| Nothing we can do to improve the environment | 0 | 27 | 5 | 1 | 16 | 0 |

**Table 19.6** Recent development in environment/natural resource management in SAVA area

| 2004 | 2005 | 2006 | 2007 | 2008 | 2009 | 2010 | 2011 | 2012 |
|---|---|---|---|---|---|---|---|---|
| Masoala National Park deforestation (ha) | | | | | | | | |
| – | – | – | 45 | – | 5.5 | 40.5 | 17 | 9.25 |
| Illegally logged trees (no.) | | | | | | | | |
| 212 | 182 | 165 | 165 | 36 | 6970 | 4163 | 853 | 91 |
| Illegally captured lemurs (no.) | | | | | | | | |
| 23 | 7 | 10 | 2 | 6 | 11 | 42 | 91 | 13 |

YNU-SLER (2013b)

The outcome of the questionnaire surveys (n=62) showed that 88 % of respondents regarded the surrounding environment as either severely degraded or degraded, and 87 % regarded the forests as severely depleted or depleted. Respondents at an income level of less than one dollar per day accounted for 70 %. The average fertility rate was in the range of 3.2–4.6 per household, which was equivalent to the national average of 4.5. With regard to education, 58 % of the respondents had completed no more than junior high school. In terms of interventions to arrest environmental degradation and promote restoration of the already-degraded environment, almost all the respondents supported the promotion of environmental education and enforcement of a penalty on illegal logging. On the other hand, only 48 % supported the idea of raising the tax on sales of charcoal, while just 19 % supported the idea of increasing the tax on land use. It showed that the options for generating additional payments on the part of local people were not favored.

It is also interesting to note the people's understanding on the causes of *lavakar* or landslides. By interviewing experts and local people, and observing sites, the groups discovered that massive landslides occur in Madagascar due to a mixture of excessive logging and deforestation (Table 19.6), and tectonic movement. It was explained that in the process of illegal logging and deforestation, the root systems of the trees degrade or disappear and create small spaces in the soil. Rainwater

**Table 19.7** Social capacity assessment—Madagascar highlights

| | Factor | | | | | |
|---|---|---|---|---|---|---|
| Actor | Policy and law | Staff | Budget | Knowledge and technology | Law enforce-ment | Capacity building |
| Government | + ↘ | – ↘ | – ↘ | + ↘ | – ↘ | – ↘ |
| Firms and Industry | + → | + → | + ↘ | ++ → | + → | + → |
| INGOs/NGOs | ++ → | + → | + ↘ | ++ → | ++ ↘ | + → |

+++: Abundant; ++: Adequate; +: Minimal; –: poor/rare; n/a data not available; ↑ Rapidly increasing; ↗ Increasing; → No change; ↘ Decreasing; ↓ Rapidly decreasing
Developed from Oo and Aung 2013, unpublished

intrudes into such spaces and makes soil structures fragile. Such processes cause *lavakar* or landslides, possibly prompted by tectonic movement. The relative weights to be given to the various interventions for arresting *lavakar* therefore depend on whether people emphasize deforestation or tectonic movement as the major cause of *lavakar*. In the questionnaire surveys, 35 respondents indicated slash and burn as a major cause of *lavakar*, followed by logging (32), heavy rain (9), gravity (7), and tectonic movement (2). This result indicates that the local people have a reasonable level of understanding with regard to the causes of *lavakar* and there is potential to create a basis for undertaking collective actions to arrest *lavakar* and restore the degraded environment.

In the interviews with experts, officials, and local stakeholders, the groups strived to collect information on social capacity for managing the environment and ecosystems. The 2009 political crisis had compelled the Government of Madagascar to resort to austere fiscal and administrative policies, particularly in the environment and forestry sectors. The budget and the number of staff for the Ministry of Forestry and Environment had been cut substantially as donor countries had suspended economic assistance, except in humanitarian areas. They had done so on the grounds that the presidential election conducted in 2009 was not consistent with constitutional procedures and was considered to be a coup d'état or unconstitutional change of government. A democratic and constitutional election was planned thereafter and the first round of the presidential election was held on October 25, 2013, with the involvement of international election monitoring. The second and final round of the presidential election is planned to be held on December 20, 2013. The process of electing a president in a manner that satisfies constitutional requirements and internationally acceptable procedures is expected to ameliorate the current international financial assistance to Madagascar. Yet, as of October 2013, donor countries had not yet restored their economic assistance to the levels seen prior to 2009, and securing the finances for environment and forestry governance remained an arduous task. With the reduction in financial inflows from overseas, public-sector, business, and civil society organizations all continue to face economic constraints.

Based on information collected during the field surveys, questionnaires, and interviews, the social capacity for environmental and ecosystem management was assessed following the actor-factor analysis approach by analyzing the levels of (1) policy and law, (2) staff, (3) budget, (4) knowledge and technology, (5) law enforcement, and (6) capacity building (Table 19.7). While a certain level of accumulated knowledge and application of technology was acknowledged, the overall trend of social capacity was at a level that was barely sufficient and was either declining or unchanging.

## 19.4 Lessons Learned and Ways Forward to Improve the Impacts of Participatory Sustainability Research in the Environmental Leadership Development Program

The field surveys were conducted very productively and efficiently within the time and resource constraints, and the outcome of the surveys attracted reasonable praise at the outcome presentation symposium. However, a number of challenges could be pointed out, and some reflections could be also noted, to enhance the impacts of participatory sustainability research in the environmental leadership development program in the future.

The field surveys were useful for understanding the local socio-economic and environmental conditions and gave the outsider visitors further ideas about possible support to improve natural resource use and promote alternative sustainable livelihoods. However, the surveys could not reach the stage of planning and undertaking consultations on possible interventions or pilot projects. This is in contrast with the SLER programme's involvement in Rikuzentakata—a city hit by the 2011 great disaster in Japan. SLER programme students, faculty members, and partners have visited Rikuzentakata four times over the period of 2011–2013. Each visit to Rikuzentakata was for 2–3 days. One pilot initiative undertaken during the visit in September 2012 was to collect sprouts of endogenous evergreen broad leaved tree species. A total of 34 students and faculty members participated in the visit and collected 477 seedlings of three tree species, namely *Persia/Machilus thumbergii* (445), *Camellia japonica* (30), and *Eurya japonica* (2). In September 2013, 27 faculty members and students returned and observed that 40 seedlings had survived over the year out of about 200 that were transplanted into the ground in October 2012. The survival rate of the collected and transplanted seedlings was estimated to be 20 %. There were an additional 280 seedlings transplanted into the garden of the public community house, but they were inadvertently all removed in the spring of 2013. In 2013, about 400 seedlings of *Persia thumbergii* and *Camellia japonica* were collected and transplanted into seedling pots. The students reported afterwards that their proactive involvement in concrete pilot initiatives such as the collection and transplantation of evergreen broad leaved tree species gave them a sense of participation in the local process, which in this case involved reconstruction and community empowerment. It also made them feel a sense of partnership with local people, and they believed that they were making useful contributions to local communities. Further details need to be elaborated elsewhere, but the students also witnessed that the local people showed more positive reactions to the initiative of producing evergreen broad leaved tree seedlings in the midst of their preoccupation with the restoration of the coastal pine woodland—the legendary scenic site in the locality that had been completely destroyed in the tsunami of 2011.

Funding is critical for undertaking effective participatory sustainability research, and not only is it needed to cover costs, but it must also be flexible enough to allow adaptive management of research activities (Mackenzie et al. 2012). Pilot initiatives need to be planned and implemented based on an appraisal of local socio-economic and environmental conditions. A broad range of possible activities and proposed procedures could be suggested, but such plans and the outcomes to be expected may not necessarily be stated rigidly as they will need to be adjusted in accordance with the findings and outcomes of local consultations.

It is vital to put in place an institution that can promote participatory sustainability action research as a part of leadership development programs in higher education. There are a number of centers and institutions established within, or in partnership with, universities to promote sustainability science research in higher education (CLiGS, CML, GMV, and the Sustainability Institute). There was one case in which an NGO was established to support sustainability action research (Harada Laboratory 2012). The operation of such centers or institutions plays an important role in establishing links between students and experts on a wide range of sustainability and environmental issues and in supporting trans-disciplinary science and its pragmatic application (MOEJ 2007). If such a center or institution could be established to support the type of activities undertaken within YNU through the SLER programme, it would be highly effective, offering more reliable continuity and adaptive management of research and operational activities.

Integration of research and education is another essential factor. A number of the educational courses can be inter-linked with each other, but the students do not necessarily take such inter-linked courses in their entirety, and the universities or faculties do not design and offer courses based on a step-by-step approach. In the case of the SLER programme, there were students who participated in the intensive course on integrated risk management and resilience and thereafter participated in the course on Asia–Africa II that mainly comprised a group study tour to Madagascar. However, not all students take both courses, although both courses employ the same approach applied in Japan and an overseas developing country. The course on local risk and resource management was an additional course that was used to follow up on the field studies undertaken in Madagascar in particular, but again the students attending the Asia–Africa II course do not necessarily attend the local risk and resource management course.

Meanwhile, further analysis is required on the linkages between individual research for the students' degree dissertations on the one hand, and group research, other course work, or leadership development program activities on the other. While research methods and approaches may be shared or related, they do not necessarily correspond to each other directly. Linking the two sides of the equation is no straightforward task, as the students have diverse interests and areas of focus even though they all address environment and sustainability issues.

Environment and sustainability education is still in its developmental stages and the supporting faculty and administrative members operate outside the university's main budget, relying on subsidies from the central government. In the case of the SLER programme, this means the Japanese Ministry of Education, Culture, Sports, Science and Technology (MEXT). It is essential to integrate such supporting staff and operational costs into the core staff and budget of the university in order to ensure continuity, up-scaling, and constructive evolution, as well as to enhance the impact of sustainability science and leadership development.

There are a number of potential and emerging opportunities for undertaking effective sustainability science and leadership development in higher education; universities and academics can do still more to create links between science, policy, and stakeholders in order to build a sustainable society. Both providers and receivers of sustainability education need to reconsider their thinking and approaches to help research and education transform themselves to increase their capacity to forge sustainability in society. Good practice and past trial experiences must be shared more widely and substantively and capitalized upon to develop effective policies, programs, curricula, institutions, and partnerships for fostering sustainability science and leadership development.



## References

Adandedjan C, Niang A (2006) Forging links between research and development in the Sahel: the missing link. In: Bessette G (ed) People, land and water – participatory development communication for natural resource management. Earthscan, London, pp 217–220

Aggrecy N, Wambugu S, Karugia J, Wanga E (2010) An investigation of the poverty-environmental degradation nexus: a case study of Katanga Basin in Uganda. Res J Environ Earth Sci 2(2):82–88

Asia Pacific Forum for Environment and Development (APFED) (2010) APFED II Final Report. http://www.apfed.net/apfed2/APFED_II_Final_Report_for_CD.pdf. Accessed 28 Sept 2013

Blackstock LD, Kelly GJ, Horsey BL (2007) Developing and applying a framework to evaluate participatory research for sustainability. Ecol Econ 60:726–742

Center for Leadership in Global Sustainability (CLiGS) (n.d.) About. http://vatech.beekeeperdev.com/about/. Accessed 5 Oct 2013

Centre for Environment and Sustainability (GMV) (n.d.) Collaboration for sustainable development. http://www.gmv.chalmers.gu.se/english/. Accessed 4 Oct 2013

Chambers R (1994a) The origins and practice of participatory rural appraisal. World Dev 22(7):953–969

Chambers R (1994b) Participatory rural appraisal (PRA): challenges, potentials and paradigm. World Dev 22(10):1437–1454

Chambers R (2007) From PRA to PLA and pluralism: practice and theory. Institute of Development Studies, Brighton

Dedeurwaerdere T (2013) Transdisciplinary sustainability science at higher education institutions: science policy tools for incremental institutional change. Sustainability 5:3783–3801

Harada Laboratory (2012) Report on the activities of PGV in Indonesia. http://www.u-hyogo.ac.jp/shse/harada/PGV_Web/indonessia.pdf. Accessed 5 Oct 2013

Institute of Environmental Sciences (CML) (n.d.) General information – overview of activities of Institute of Environmental Sciences. http://www.cml.leiden.edu/organisation/about/aboutcml.html. Accessed 4 Oct 2013

Kelvin SHP, Balmoford A, Bradbury RB, Brown C, Butchart SHM, Hughes FMR, Stattersfield A, Thomas DHL, Walpole M, Bayliss J, Gowing D, Jones PGJ, Lewis SL, Mulligan M, Pandeya B, Stratford C, Thompson JR, Turner K, Vira B, Willcock S, Birch JC (2013) TESSA: a toolkit for rapid assessment of ecosystem services at sites of biodiversity conservation importance. Ecosyst Serv 5:e51–e57

Kobayashi M (2012) Forging policy and institutional frameworks to promote access to environmental information. In: Institute for Global Environmental Strategies (IGES) (ed) Institute for global environmental strategies. IGES, Hayama, pp 34–53

Mackenzie J, Tan PL, Hoverman S, Baldwin C (2012) The value and limitations of participatory action research methodology. J Hydrol 474:11–21

Matsuoka S, Murakami K, Aoyama N, Takahashi Y, Tanaka K (2008) Capacity development and social capacity assessment (SCA)

Millennium Ecosystem Assessment (2005) Ecosystems and human well-being: synthesis. Island Press, Washington, DC

Ministry of Environment, Japan (MOEJ) (2007) On-going overseas research. Committee on Environmental Leadership Development in Higher Education for a Sustainable Asia. http://www.env.go.jp/council/34asia-univ/y340-04/mat07.pdf. Accessed 3 Oct 2013

Ostrom E (1990) Governing the commons. Cambridge University Press, Cambridge

Ostrom E (2000) Collective action and the evolution of social norms. J Econ Perspect 14(3): 137–158

Petkova E, Maurer C, Henninger N, Irwin F, Coyle J, Hoff G (2002) Closing the Gap: information, participation and justice in decision-making for the environment. World Resource Institute, Washington, DC

Rockström J, Steffen W, Noone K, Persson K, Chapin FS III, Lambin E, Lenton TM, Scheffer M, Folke C, Schellnhuber H, Nykvist B, De Wit CA, Hughes T, van der Leeuw S, Rodhe H, Sölin S, Snyder PK, Costanza R, Svedin U, Falkenmark M, Karlberg L, Corell RW, Fabry VJ, Hansen J, Walker B, Liverman D, Richardson K, Crutzen P, Foley J (2009) Planetary boundaries: exploring the safe operating space for humanity. Ecol Soc 14(2):32

Sow F, Adjibade A (2006) Experimenting with participatory development communication in West Africa. In: Bessette G (ed) People, land and water – participatory development communication for natural resource management. Earthscan, London, pp 153–157

Sustainability Institute (n.d.) Overview. http://www.sustainabilityinstitute.net/about/overview. Accessed 4 Oct 2013

Takeuchi K (2013) Science and innovation for a sustainable future: the role of sustainability science. International symposium on developing leaders for managing risks and promoting sustainability toward establishing a resilient and sustainable society, Tokyo, 23 September 2013. http://www.sler.ynu.ac.jp/node/592. Accessed 23 Sept 2013

Thermo Fisher Scientific (n.d.) Gardening & Soil pH. www.eutechinst.com/tips/ph/15_soil_ph.pdf. Accessed 2 Oct 2013

United Nations General Assembly (UNGA) (1992), A/CONF.151/26 (Vol. I). Report of the United Nations conference on environment and development, 12 August 1992

United Nations General Assembly (UNGA) (2012). Resolution 66/288. The future we want, 11 September 2012

World Resource Institute (WRI) (2013) Jakarta declaration for strengthening the right to environmental information for people and the environment, 1 May 2013. http://pdf.wri.org/jakarta_declaration_for_strengthening_right_to_environmental_information.pdf

Yasunari T (2013) Future earth – a new international initiative toward global sustainability. International symposium on developing leaders for managing risks and promoting sustainability toward establishing a resilient and sustainable society, Tokyo, 23 September 2013. http://www.sler.ynu.ac.jp/node/592. Accessed 23 Sept 2013

Yokohama National University-Leadership Development Programme For Sustainable Living With Environmental Risks (YNU-SLER) (2013a) Newsletter, Vol. 6, March 2013. http://www.sler.ynu.ac.jp/en/newsletter. Accessed 1 Oct 2013

Yokohama National University-Leadership Development Programme for Sustainable Living with Environmental Risks (YNU-SLER) (2013b) Report – Asia–Africa II Group Study Tour to Madagascar https://sites.google.com/site/ynuslermadagascar/final-report. Accessed 28 Oct 2013

# Chapter 20
# Rural Landscape Conservation in Japan: Lessons from the *Satoyama* Conservation Program in Kanagawa Prefecture

**Osamu Koike**

**Abstract** Japanese call rural landscapes '*satoyama.*' '*Sato*' means village and '*yama'* means hill or forest. *Satoyama* in the past produced much of the food, wood for fuel, timber, and water for communities. However many *satoyama* have rapidly deteriorated due to industrialization and urbanization. It was in the 1990s that people in general and scientists began to recognize the multiple benefits of *satoyama* landscapes in Japan. This led to the proliferation of *satoyama* conservation groups across Japan at the turn of the century. However, it is difficult for local action groups to rehabilitate devastated farmlands and forests through their own efforts alone. It requires policy measures to encourage citizen engagement in *satoyama* conservation programs. In this paper the author addresses governance issues in rural landscape conservation, referring to the case of the *satoyama* conservation program in the Kanagawa region.

**Keywords** Citizen engagement • Local government • Rural landscape • *Satoyama*

## 20.1 Crisis of *Satoyama* Landscapes in Japan

Japanese people call rural landscapes '*satoyama.*' '*Sato*' means village and '*yama*' means hill or forest. Takeuchi (2003) defines '*satoyama*' as the 'secondary woodlands and grasslands adjacent to human settlements.' Farmers in *satoyama* areas have been living in harmony with nature for centuries, and such areas make the landscapes of rural communities multi-functional. 'Landscape' is defined as 'an area, as perceived by people, whose character is the result of the action and interaction of natural and/or human factors,' according to the European Landscape Convention. *Satoyama*, meanwhile, is defined by a study team from the United Nations University as a

O. Koike (✉)
Graduate School of International Social Sciences,
Yokohama National University, Yokohama, Japan
e-mail: okoike@ynu.ac.jp

N. Kaneko et al. (eds.), *Sustainable Living with Environmental Risks*,
DOI 10.1007/978-4-431-54804-1_20, 

Japanese term for landscapes that comprise a 'mosaic of different ecosystem types including secondary forest, agricultural lands, irrigation ponds, and grasslands, along with human settlements' (Duraiappah and Nakamura 2012: 3). For a long time, Japanese people have considered *satoyama* landscapes to be a symbol of the traditional Japanese lifestyle and a spiritual home for the population.

In Japan, *satoyama* occupy nearly 40 % of the total landmass. *Satoyama* in the past produced much of the food, wood for fuel, timber, and water for communities. However many *satoyama* have deteriorated rapidly due to industrialization and urbanization. Farmers use chemical fertilizers instead of natural composts; people use electricity and gas instead of firewood. A decline in the economic value of agriculture and forest products has accelerated rural–urban migration of younger generations and the aging of populations in rural areas (Duraiappah and Nakamura 2012:5). In the urban fringes, housing development encroaches on the rural landscapes, which results in deterioration of the ecosystem and biodiversity in these areas.

It was in the 1990s that people in general and scientists began to recognize the multiple benefits of *satoyama* landscapes in Japan (Takeuchi 2003). The 'slow food' or 'slow life' movement inside and outside the country had partially contributed to the reappraisal of *satoyama* in terms of food security, therapy, and environmental education. It was biologists that emphasized the rich biological diversity of *satoyama* areas. This led to the proliferation of *satoyama* conservation groups across Japan at the turn of the century.

## 20.2 Citizens' Actions for *Satoyama* Conservation

In environmental conservation, the role of government has been widely recognized based on the discussion presented in the 'Tragedy of the Commons' (Hardin 1968). Controlling greenhouse gas emissions, preserving marine resources such as whales and coral reefs, and establishing national parks for wild animals are all examples of government intervention. In rural landscape conservation, however, smaller populations mean that political action is not guaranteed, even though rural landscapes contain valuable resources such as beautiful scenery, cultural heritages, and biodiversity (Swaffield and Primdahl 2010). It requires a strong political commitment to take policy measures for conserving rural landscapes, such as land-use regulation, direct income support for working farmers, and allocation of budgets for non-agricultural functions including the conservation of historical and cultural heritages. If no positive action is taken, it will result in the deterioration of rural landscapes in the near future.

In Japan, citizens facing the decline of rural landscapes have taken action to regenerate abandoned farmlands and forests before government intervention. As agricultural land and forest in rural areas are private property, national and local governments find it difficult to provide direct support for rural communities. For better or for worse, this encourages collaborative actions between farmers and community members for the conservation of rural landscapes. Such collaboration is deeply rooted in the historical heritage of rural governance in Japan.

In the pre-modernization era, farmers in rural villages developed a system of 'common property' for the common use of water, grass, wood, and timber in their communities. This type of 'common pool of property' is called '*iriaichi*' in Japanese. Margaret McKean (1992) notes that 'well organized communities of co-owners are capable of protecting and managing their property quite well' (McKean 1992, p. 253). Even though the system of *iriaichi* has disappeared in contemporary Japan, rural community members keep a mindset of 'mutual help' as in the past. It enables the lending and borrowing of abandoned farmlands among the community members outside of traditional property ownership. In the process of landscape conservation, urban citizens have also joined in the conservation efforts, organizing voluntary action groups. As a result, various patterns of 'collaboration' or 'partnership' have developed among farmers and civic organizations across the country.

## 20.3 Governmental Policies for *Satoyama* Conservation

It was in the mid-90s that a renewed and growing interest revived the traditional rural landscapes of *satoyama* in Japan (Duraiappah and Nakamura 2012, p. 2). A trigger was the Earth Summit in Rio de Janeiro of 1992 that discussed sustainable development in the world. After the Rio Declaration, Japanese national and local governments began looking at policy for rural areas in terms of environmental conservation. However policy actions remained superficial and piecemeal. In 1994, the national government established the Basic Environment Plan and referred to the need for *satoyama* conservation officially. It described the *satoyama* landscape as an area featuring substantial secondary nature where wild animals and humans can live together. It stated, 'the natural environment of this area is created through human interventions such as farming and forestry, and it is what Japanese people have long imagined as their idyllic landscape. In order to conserve the coppices, rice paddies, and irrigation ponds of *satoyama* landscapes, some action by citizens and public support are necessary' (Takeuchi 2003, p. 11).

In the National Biodiversity Strategy of 1995, the government also made the conservation of ecosystems in Japan a priority. However the strategy did not contain the term *satoyama*. It was the New National Biodiversity Strategy of 2003 that defined *satoyama* as an ecosystem that preserves biological diversity. Based on the new strategy, the Ministry of Environment (MOE) launched the *Satoyama* Conservation and Rehabilitation Model Program in 2004. In 2008, the government enacted the Basic Law on Biodiversity that defined *satoyama* areas eligible for biodiversity conservation. In 2010 the Ministry of Environment unveiled the *Satoyama* Conservation and Utilization Action Plan to provide a guideline for local government on how to conserve *satoyama* areas. It is the Ministry of Environment that has been the most dynamic actor in the conservation of *satoyama* areas. The Ministry advocates the '*Satoyama* Initiative' in international meetings on biodiversity, based on the experiences in Japan. Finally the government of Japan officially called for the promotion of the '*Satoyama* Initiative' at the COP 10 biodiversity meeting in Nagoya in October 2010 (Takahashi 2012).

Later the Ministry of Agriculture, Forestry and Fisheries (MAFF) joined the *satoyama* program. The Basic Law on Food, Agriculture, and Rural Areas of 1999 defined the multiple functions of farmlands, such as maintaining biodiversity and disaster prevention. However achievement of these newly defined functions was not initially linked to the conservation of *satoyama*. Instead MAFF introduced a system of direct income support payments for farmers operating their farms as businesses to emphasize the multi-functionality of agriculture as in the EU. MAFF believed that a direct payment would motivate farmers and lead to a reduction in the number of abandoned cultivated lands in the hilly and mountainous areas (MAFF 2011). On the other hand, the Ministry of Land, Infrastructure, Transport and Tourism (MLIT) used *satoyama* to reintegrate agricultural space into urban areas so as to establish multifunctional green spaces (Yokohari et al. 2010). Meanwhile, the Ministry of Internal Affairs and Communications (MIC) also joined the race. MIC assisted local governments to promote rural development programs in *satoyama* areas, such as eco-tourism. However there is no single agency beyond the ministries mentioned above that is responsible for integrated *satoyama* conservation programs (Takahashi 2012). This has resulted in superficial and fragmented public programs for *satoyama* conservation in national government.

Lack of positive, integrated policy measures on *satoyama* conservation at national level pushed the local action groups toward direct collaboration with local government. In the first decade of the 2000s, municipal and prefectural governments decided to enact local bylaws to conserve rural landscapes. In 2000, the City Government of Kouchi enacted a local bylaw that aimed to conserve *satoyama* areas in Kouchi city. It designated *satoyama* conservation areas for disaster relief to protect citizens from tsunami tidal waves. In 2003, Chiba Prefectural Government enacted the *Satoyama* Bylaw to support citizens' voluntary action to conserve community forests in the Chiba region. It was followed by Kanagawa Prefecture in 2007. Kanagawa's *Satoyama* Conservation Bylaw aims to rehabilitate, conserve, and develop *satoyama* areas through the collaboration of farmers, landowners, local action groups and municipal and prefectural governments.

In 2011–2013, the author conducted research on the *satoyama* conservation programs of the Kanagawa prefecture government. It revealed a number of both achievements and challenges in the conservation of rural landscapes in the Kanagawa region.

## 20.4 *Satoyama* Conservation in Kanagawa Prefecture

Kanagawa prefecture is located to the south of the Tokyo Metropolis and, with neighboring prefectures, makes up the National Capital Region. The eastern part of Kanagawa, which includes Yokohama City and Kawasaki City, is heavily urbanized. The western part, on the other hand, is mountainous and rich in nature. Rural *satoyama* areas have been protected from development, being located in the hinterlands, or comprising part of Quasi-National Park Reserved Areas. These areas are designated either as Urbanization-Restricted Areas or Agriculture Promotion Areas. However farmlands and forests in the Kanagawa region are rapidly decreasing due to urbanization and industrial development. In the areas closest to the urban districts the encroachment of housing and manufacturing damages the landscape of rural areas. Inside *satoyama* areas aging farmers give up farming and forestry and their children move away to work in offices. This results in the devastation of rural landscapes. To save *satoyama* areas from this deterioration the Kanagawa Prefectural Government enacted a bylaw in 2007, and this was followed by the *Satoyama* Conservation Action Plan in 2009.

The *Satoyama* Conservation Bylaw aims to enrich people's wellbeing by developing the multiple functions of *satoyama* in Kanagawa. *Satoyama* have a multifaceted significance, providing seasonal scenery, biodiversity, cultural heritage, fresh air, safe food, disaster relief, etc. The bylaw underscores the role of community and municipal government in the conservation of *satoyama* landscapes. It is the prefectural government that selects '*Satoyama* Conservation Areas' and accredits '*Satoyama* Conservation Partnership Agreements' between landowners and local action groups. The prefectural government provides financial and technical assistance for local action groups, who purchase grass cutters and chainsaws using subsidies. The Agriculture Division of Kanagawa Prefectural Government hosts programs to raise public awareness of *satoyama* among people of different generations, including farming experience programs for school children, *satoyama* symposiums, and other initiatives.

In the Action Plan the prefectural government set the goal of selecting 16 conservation areas and 20 partnership agreements by 2013. By 2012 13 areas had been designated as *Satoyama* conservation sites and 14 local action groups had established *Satoyama* Conservation Partnership Agreements to engage in *satoyama* conservation activities (see Fig. 20.1). They range from the smallest *Ishikawa-Maruyama* area (12 ha) to the largest *Kuno* area (2,800 ha). The larger areas include mountain forests, while farmlands are relatively modest in size and divided into small plots. Some mountain forests are the common property of landowners.

Based on the established partnership agreements, action groups borrow the abandoned farmlands and forest from the landowners to rehabilitate them for use (see Fig. 20.2). Besides the rehabilitation of farmlands and forest, local action groups have utilized the resources of *satoyama* landscapes in different ways as shown in Table 20.1. The most popular activity is environmental education. Local action groups invite school teachers and students to their *satoyama* areas to provide a 'field' for the study of nature and biodiversity. There is a growing interest in a 'hands-on' approach to environmental education among school teachers, and *satoyama* offer them the ideal opportunity for this. In some areas action groups hold various events with the local neighborhood organizations to teach traditional knowledge and cultural heritage such as traditional cooking, toys, and crafts to small children. In the *Kuno* area of Odawara City the action group invites children and their parents living in the city center to plant soba (buckwheat) as shown in Fig. 20.3. After planting the soba a member of the action group plays educational nature games with the children to teach them the various species of insects.

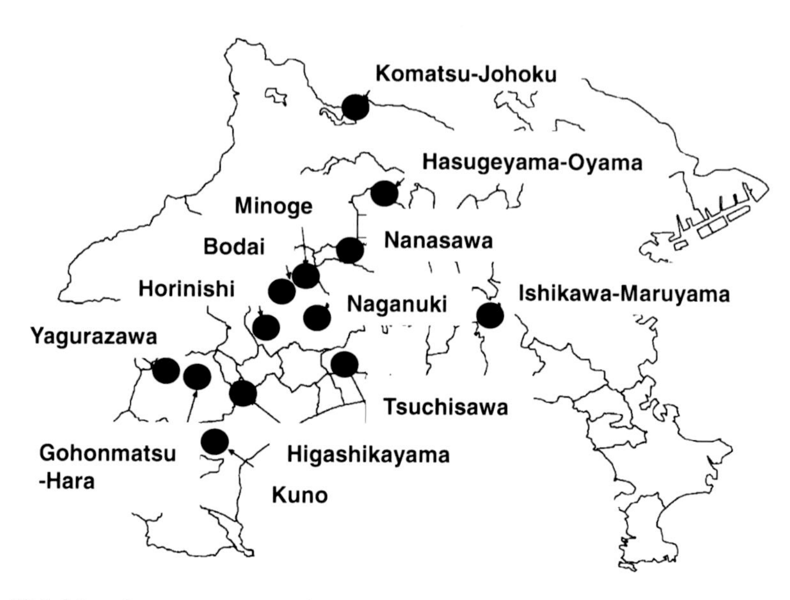


**Fig. 20.1** Map of *satoyama* conservation areas (as of 2012)

**Fig. 20.2** A rehabilitated terraced rice field in *Naganuki* (Hadano City)

Secondly, members of action groups know that *satoyama* are often used for recreation. *Satoyama* provide opportunities for such pursuits as walking, hiking, birdwatching, camping, and fishing. Some action groups work to provide facilities such as benches, signboards, and parking lots to welcome visitors. A group in

**Table 20.1** Major activities in the *satoyama* conservation areas

| Area | Major activities |
|---|---|
| *Bodai* | Rehabilitation of farmland (terraced rice fields) and forest; food education (cooking local foods); farming experience program for urban consumers in collaboration with the COOP Kanagawa |
| *Gohonmatsu-Hara* | Rehabilitation of farmland and forest; education activities for sustainable development; chrysanthemum festival |
| *Higashikayama* | Rehabilitation of farmland and forest; farming experience program; colza flower festival; biodiversity survey; social education |
| *Horinishi* | Rehabilitation of irrigation, farmlands, and forest; rice farming experience program |
| *Ishikawa-Maruyama* | Preservation of fireflies; conservation of farmlands in hilly terrain; biodiversity survey in collaboration with elementary school |
| *Komatsu-Johoku* | Woodlands conservation; preservation of fireflies; farming experience program; ancient route restoration; cosmos festival |
| *Kuno* | Rehabilitation of farmland and forest; biodiversity survey; farming (soba planting) experience program; mountain trails rehabilitation |
| *Minoge* | Rehabilitation of farmland and forest; farming experience program |
| *Naganuki* | Rehabilitation of farmland (terraced rice fields) and forest; fruit farming experience program; cooking class using local foods |
| *Nanasawa* | Rehabilitation of farmland and forest; landscape (scenery) conservation; charcoal production; micro-hydroelectric power generation |
| *Tsuchisawa* | Rehabilitation of irrigation; conservation of woodlands; biodiversity survey in collaboration with elementary school and university; cooking class; environmental education; provision of playgrounds for children |
| *Yagurazawa* | Rehabilitation of farmland and forest; promotion of nature education; chrysanthemum festival; health promotion |

**Fig. 20.3** Seeding of soba by children in *Kuno* (Odawara City)

*Komatsu-Johoku* restored ancient trails for tourists; another group in *Kuno* established a trail up Mt. *Myojo*. Several groups host flower festivals and cultural events for visitors from urban areas. These actions are sustained by the emotion that the beautiful scenery of *satoyama* inspires in the local residents; they are the pride of the community. Therefore action groups plant trees and restore devastated terraced rice fields to rehabilitate landscapes.

The third facet of *satoyama* is health and welfare; walking around *satoyama* cleanses body and soul. Farming and forestry are beneficial for depression, and there is a growing interest in 'agri-therapy' and 'forest therapy' among health policy professionals. Consequently, some action groups use farmlands and forest to promote health and welfare. In the *Yagurazawa* area of Minami-Ashigara City, the local action group invites patients of a mental hospital to engage in farming activities.

The most striking finding is that a variety of conservation activities are 'home-grown.' Municipal and prefectural governments provide only limited subsidies and technical advice for action groups; if necessary, action groups organize seminars and invite lecturers from outside. It is notable that a horizontal, regional network of *satoyama* conservation has developed in Kanagawa prefecture.

## 20.5 The Way Forward

Of course it is too early to evaluate the outputs and outcome of *satoyama* conservation programs in Kanagawa. The rehabilitation and conservation of *satoyama* areas in Kanagawa remain in their early stages at best. Accordingly it is better to discuss the challenges for the way forward.

First, Japan's aging population and dwindling younger generation are quite serious issues in *satoyama* communities. In mountainous areas farmers face the problem of increasing wild boar and deer populations that eat agricultural products, and it is difficult for the senior citizens to protect farmlands from these wild animals. It requires the government to take positive policy measures to involve younger generations in the conservation of *satoyama* landscapes.

A second challenge is the fact that intergovernmental cooperation is necessary for successful *satoyama* conservation. Vertically segmented public administration and fragmented programs impede integrated community actions for conservation. In this regard the role of prefectural government is quite important. As individual municipal governments lack incentives to cooperate, it is the prefectural government that can initiate the regional cooperation of municipalities.

The third is financial support for local action groups. Most farmers in the *satoyama* areas do not engage in for-profit agricultural production any more. Their agriculture is more a kind of self-sufficiency. They produce rice and vegetables for the consumption of their families, and feel joy in welcoming children to the farmland to teach them how to live with nature. Therefore, rather than seeking direct payments for income support, they want local government to reform legal and institutional frameworks to make their voluntary actions more effective. For example, under current land use regulations it is difficult for them to construct a rest room at the corner of farmland. Needless to say, it is important to develop measures to raise funds for these positive activities, and it will therefore be possible for local action groups to engage in business to acquire financial resources. In the process private companies will be able to engage in the conservation of forests in order to demonstrate their 'corporate social responsibility' (CSR). In this case, local governments should introduce strategic measures such as 'carbon offset credit' to attract private enterprises to contribute to *satoyama* conservation.

Finally, both national and local governments should seek international cooperation on *satoyama* conservation. It is possible to link the practices of *satoyama* conservation with the Education for Sustainable Development (ESD) project by the United Nations University/UNESCO and the '*Satoyama* Initiative' promoted by the United Nations. Furthermore, it is highly recommended for Japanese local governments to work with European local governments that promote landscape conservation under the European Landscape Convention of 2000 (Dejeant-Pons 2011).

## 20.6 Concluding Remarks

Farmlands and forests in *satoyama* areas are individually owned private properties. However, faced with the issues of an aging population and an increase in abandoned farmlands and forests, conservation groups organize to rehabilitate their *satoyama* collectively. The groups, comprising farmers, landowners, and community members, do so not only to promote their personal interests but for the well-being of future generations. In the past, Japanese rural communities had a system of common property management named '*iriaichi*.' The *iriaichi* system has disappeared from Japan in the process of agricultural modernization and land reform, and has led to an increase in farms operated as businesses. However, the challenge now facing *satoyama* conservation is not to attempt to revive this old system of common property management. As Kanagawa's *satoyama* conservation programs demonstrate, the orientation is instead toward a new mode of community-based conservation based on collaboration by landowners and local action groups. The bylaws encourage common use of privately owned farmlands for the rehabilitation and conservation of rural landscapes.

It should be noted that both community members and action groups work together to keep these landscapes beautiful for future generations. They do not demand an income support program from the government, rather they require the revision of national laws to make the best use of their properties. Current agricultural policy is premised on promoting agricultural industrialization and 'commercialization.' However most farmers in the *satoyama* areas are not engaged in farming as a business and their main source of income is not from the farming. They are simply individuals who like the *satoyama* lifestyle and living with nature.

To support the activities of local action groups in *satoyama* areas it seems necessary to create a strong network of professionals. In the case of environmental education, for example, school teachers and the parents of school children would welcome university professors to provide scientific knowledge on the subject. Meanwhile, where health promotion is concerned, a network of medical doctors, schools of medicine, public health centers, and hospitals would enhance the effective use of *satoyama* properties for public health promotion.

From a sociological perspective, civic engagement in *satoyama* conservation will enhance the so-called 'social capital' of urban populations (Putnam 2000). Working together with farmers, urban dwellers can feel and understand the meaning of community ties and 'quality of life' in a material society.

This leads us back to the debate on the responsibility of government in the provision of public goods. People in *satoyama* think that these landscapes are public goods necessary for the well-being of mankind in terms of 'safety and security.' The author expects a new mode of civic governance will develop through a combination of community-based conservation and local government policy measures to develop the multi-functionality of *satoyama* landscapes. It will take on a different shape from western countries due to the influence of traditional values inherent in the sustainable lifestyle of Japan.



## References

Dejeant-Pons M (2011) The European landscape convention: from concepts to rights. In: Egoz S, Makhzoumi J, Pungetti G (eds) The right of landscape: contesting landscape and human rights. Ashgate Publishing, Burlington

Duraiappah AK, Nakamura K (2012) The Japan satoyama satoumi assessment: objectives, focus and approach. In: Duraiappah AK et al (eds) Satoyama-satoumi ecosystems and human well-being: socio-ecological production landscapes of Japan. United Nations University Press, Tokyo

Hardin G (1968) The tragedy of the commons. Science 162:1243–1248

Ministry of Agriculture, Forestry and Fisheries (MAFF) (2011) FY 2011 annual report on food, agriculture, and rural areas in Japan: summary. http://www.maff.go.jp/j/wpaper/w_maff/h23/pdf/e_all.pdf. Accessed 29 Aug 2013

McKean M (1992) Success on the commons: a comparative examination of institutions for common property resource management. J Theor Polit 4(3):247–281

Putnam RD (2000) Bowling alone: collapse and revival of American community. Simon and Schuster Paperbacks, New York

Swaffield S, Primdahl J (2010) Globalisation and local agricultural landscapes: patterns of change, policy dilemmas and research questions. In: Primdahl J, Swaffield S (eds) Globalisation and agricultural landscapes: changing patterns and policy trends in developed countries. Cambridge University Press, Cambridge

Takahashi T (2012) What and how effective have the main responses to address changes in satoyama and satoumi been? In: Duraiappah AK et al (eds) Satoyama-satoumi ecosystems and human well-being: socio-ecological production landscapes of Japan. United Nations University Press, Tokyo

Takeuchi K (2003) Satoyama landscapes as managed nature. In: Takeuchi K et al (eds) Satoyama: the traditional rural landscape of Japan. Springer, Tokyo

Yokohari M, Amatim M, Bolthouse J, Kurita H (2010) Restoring agricultural landscapes in shrinking cities: re-inventing traditional concepts in Japanese planning. In: Primdahl J, Swaffield S (eds) Globalisation and agricultural landscapes: changing patterns and policy trends in developed countries. Cambridge University Press, Cambridge

# Chapter 21
# Enhancing Students' Ecological Thinking to Improve Understanding of Environmental Risk

**Norizan Esa, Hashimah Yunus, Nooraida Yakob, Mahamad Hakimi Ibrahim, and Mardiana Idayu Ahmad**

**Abstract** The current rate of human development poses a major threat to the ecological balance of the environment, and this is exacerbated by climate change effects. It is therefore important for people to understand the ecology of different locations so they can make informed decisions that will not have an adverse effect on the environment. In this regard, students form a significant group for whom sound ecological thinking is necessary as they will be the future leaders and decision makers who can ensure the world's continued sustainability. This chapter presents the results of an ecological education project that successfully expanded students' ecological thinking. Evidence of this change is taken from interview responses before and after the ecological education project. The interview was based on a set of six photographs of different situations found in the environment. Responses were analyzed based on an ecological thinking framework developed by the researchers. This framework consisted of two components: understanding of ecology concepts and understanding of the impacts of human activity on ecosystems. The ecological education project succeeded in increasing students' ecological thinking, thereby increasing their awareness of environmental risks.

**Keywords** Ecological education • Ecological thinking • Environmental risk awareness

N. Esa (✉) • H. Yunus • N. Yakob
School of Educational Studies, Universiti Sains Malaysia, Penang 11800, Malaysia
e-mail: norizanesa@usm.my; myshima@usm.my; nooraida@usm.my

M.H. Ibrahim • M.I. Ahmad
School of Industrial Technology, Universiti Sains Malaysia, Penang 11800, Malaysia
e-mail: mhakimi@usm.my; mardianaidayu@usm.my

N. Kaneko et al. (eds.), *Sustainable Living with Environmental Risks*,
DOI 10.1007/978-4-431-54804-1_21, 

## 21.1 Introduction

Of all influences on the global environment, human activities have exerted the greatest impact. The rapid pace of development means that the ecological balance of the environment is faced with enormous threats. Among these, the top five pressures on biodiversity globally are the loss, alteration, and fragmentation of habitats; overexploitation of wild species populations; pollution; climate change; and the introduction of invasive species. These pressures were evidenced by a 28 % decline in the global Living Planet Index between 1970 and 2008. The decline subsequently continued, reaching 30 % in 2012 (WWF 2012). The Living Planet Index is a measure of global biodiversity change based on the world's vertebrate population.

Meanwhile, the Global Ecological Footprint—a measure of the amount of natural resources consumed globally against the world's biocapacity—increased continuously. In 2008, the figure was 18.2 billion global hectares (gha). This amounted to 18.2 billion hectares of land to supply the resources necessary to fulfil lifestyle needs and absorb waste for every person on Earth. This means that the Earth requires 1.5 years to regenerate the natural resources that global consumption uses in 1 year (McRae et al. 2008). It is thus necessary to restore, conserve, and protect natural ecosystems and biodiversity so that biological productivity and ecosystem services can be maintained (WWF 2002). This includes preserving the world's biodiversity and reducing the impact of human activity on natural habitats (WWF 2008a, b). However, realization of these goals requires public support and participation. It is therefore important for people to understand the ecology of different locations so they can make informed decisions that will not have an adverse effect on the environment. Students form a significant group for whom sound ecological thinking is necessary as they will be the future leaders and decision makers who can ensure the world's continued sustainability.

## 21.2 Ecological Thinking

Balgopal and Wallace (2009) reported Berkowitz's definition of ecological thinking as a combination of ecological understanding and environmental awareness. Ecological understanding refers to understanding of the general concepts in ecology. This includes food webs, trophic levels, carrying capacity, and population dynamics. When people acquire ecological understanding, they tend to also consider their position and role in the ecosystem (Orr 1992, and van Weelie 2002 in Balgopal and Wallace 2009). The definition is further extended to include understanding the impact of human activity on the ecosystem through recognition and application of ecology concepts. This understanding is referred to as ecological literacy (Balgopal and Wallace 2009) and is described as being on a continuum. At one end is the ability to identify dilemmas and propose decisions together with their consequences. This ability diminishes progressively toward the other end of the continuum where there is insufficient understanding to explain how human action impacts on the ecosystem.

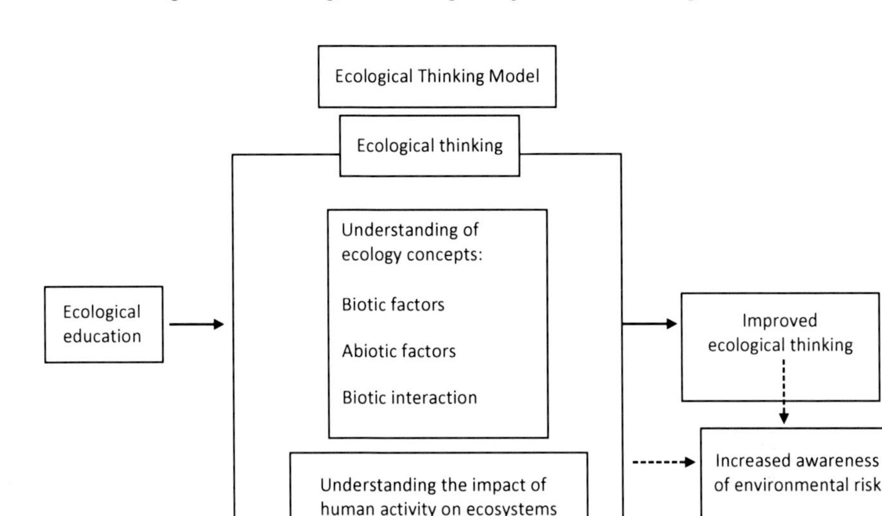


**Fig. 21.1** Conceptual framework of ecological education

The researchers have developed a conceptual framework (Fig. 21.1) based on the definitions given. Ecological thinking involves understanding concepts in ecology including biotic factors, abiotic factors, and biotic interaction. It is complemented by understanding the impact of human activity on ecosystems. Ecological thinking can be improved through ecological education. Improvement of ecological thinking, and understanding the impact of human activity on ecosystems, will eventually lead to increased awareness regarding environmental risk. An ecological education project was therefore conducted with secondary school students as subjects to effect an improvement in their ecological thinking.

## 21.3 Case Study

Changes in students' ecological thinking were investigated. The changes were facilitated by an ecological education project carried out with secondary school students as subjects. In this project, students set up themed organic gardens. They worked in groups and each group created two gardens: a wild garden and a garden on a specific theme chosen by the group. In the wild gardens, plants in the designated area were left to grow freely and no additional fertilizer was applied. In the themed gardens, students planted species they identified as suitable for their chosen theme. This project was conducted for 3 months. During this time, the students collected vermicompost produced by earthworms they reared on cow dung and food waste. They also carried out experiments to compare the effect of using either organic fertilizer in the form of vermicompost, or chemical fertilizer, on the plants in their themed gardens. In addition, they recorded their observations of other organisms found in their gardens.

Data with regard to the students' ecological thinking were collected using the photo-elicitation interview technique. This is a technique often used in social science by anthropologists and sociologists (Hurworth 2003), as well as in psychology and education, albeit minimally in the latter cases (Harper 2002). Apart from being user friendly and requiring only simple technology to produce, photographs can be used either on their own as content for discussion or as a part of the overall interview process (by varying the way they are presented). Such use enables the interviewer to probe responses about social relationships (Epstein et al. 2006). Furthermore, photo-elicitation incorporates visual language with verbal language (Hurworth 2003) and both interviewer and interviewee share the same visualization that becomes the focus of the interview. Absence of such images requires both parties to conjure their own image of the subject in their minds. In this case congruency of the visualization cannot be ensured as both parties arrive with different experience and prior knowledge.

A total of 140 students aged between 15 and 16 years participated in the project. A sample of students were interviewed prior to its start (pre-test); then they were again interviewed at the end (post-test). Using the photo-elicitation technique, photographs were shown to the students and questions posed to elicit responses (Epstein et al. 2006; Hurworth 2003). The photographs constituted six images of the environment in various situations, as depicted in Fig. 21.2. They included a pristine rainforest, a paddy field, residential apartments, a hill slope being cleared, a riverside settlement, and chemical spraying. Based on the ecological thinking model, the interview covered two aspects, namely understanding of ecological concepts and understanding of the impact of human activity on ecosystems. The students' responses were probed to gain more information about their thinking with regard to the situations presented in the photographs and their responses. After students were interviewed about the first photograph, they were then interviewed about the second photograph, and this process was repeated until all six photographs had been covered. The same process was followed for both the pre-test and the post-test. Interview data from six students are presented in this case study. Comparison of the pre-test and post-test interview responses is made to identify the changes in their ecological thinking after participating in the ecological education project. The interviews were transcribed and analyzed to extract data relevant to the components of the ecological thinking model. Table 21.1 gives a summary of the findings.

## 21.4 Conclusion

The ecological education project succeeded in improving students' ecological thinking. Their understanding of basic concepts in ecology improved. More importantly, students became more aware of the threats to the environment posed by human activity.

Photo 1 Forest and river

Photo 2 Paddy field

Photo 3 Residential apartments

Photo 4 Hill slope clearing

Photo 5 Riverside settlement

Photo 6 Herbicide spraying

**Fig. 21.2** Set of photographs for interview

Education is a tool that can enhance understanding of ecosystems in terms of ecological concepts and the effects of human activity on ecosystems. The project undertaken in the case study is one such example. However, there is potential for further research into the ecological thinking of students to assess differences between them with respect to certain demographic factors. Factors for consideration could include their own experience of natural ecosystems, the type of residential area in which they live—rural or urban, for example, or located near a nature reserve or forest—and their worldview.

**Table 21.1** Ecological thinking of students

| Themes | Pre-test | Post-test | Conclusion |
|---|---|---|---|
| Biotic factors | Students had limited knowledge and ideas relating to biotic factors | Students' knowledge of biotic factors had increased in terms of the examples they gave. Furthermore, students were able to elaborate and explain their examples | Students' knowledge of biotic factors had increased due to the educational intervention provided. Students had become more aware of the biotic factors in our environment |
| | Example of response—Respondent 6: *There are wild plants, plants on the mountain, a river, and creeping wild plants* | Example of response—Respondent 1: *Near the plants behind (in the photograph) maybe there are monkeys, ant eaters, and other animals that live in the forest. Near the planted trees there will be mudskippers, mice, and ants. In the water, there will be fish* | |
| Biotic interaction | Students' ideas relating to biotic interaction were mainly concerned with a simple food chain. | Students' ideas and knowledge relating to the food chain had become more sophisticated. The food chain they described had been extended to a food web and included higher trophic levels. | Students' understanding and knowledge of biotic interaction had increased. They were able to explain and describe biotic interaction with improved knowledge of food sources, food webs, and trophic levels. |
| | Example of response—Respondent 1: *Tigers feed on mouse deer or mice* | Example of response—Respondent 6: *OK, I see a river; in the river there might be fish; maybe the fish can feed on the seaweed and the small fish will be eaten by the bigger fish, which will then be eaten by the crocodiles. OK, trees, for example, trees have fruits and the fruits might be eaten by the squirrels or monkeys. Both squirrels and monkeys might be hunted by humans* | |

| | | | |
|---|---|---|---|
| Abiotic factors | Students' knowledge and ideas relating to abiotic factors were simplistic and limited<br><br>Example of response—Respondent 6:<br>*Housing area* | Students were able to provide a greater variety of examples. They could also provide explanations and elaborate further on the examples given. This implies an increase in their knowledge of abiotic factors<br><br>Example of response—Respondent 6:<br>*First, the river is the source of water for the living animals. Second, in the water cycle, water from the river flows to the sea. Water evaporates from the surface of the sea to form clouds. So when there is cloud it will not be too hot. The river can also be the medium for transportation.*<br>*The forest is important because if there is no forest...now that we have many logging activities taking place, the ozone layer is becoming thin...so the forest is important for...one reason is for the animal habitat, and in the forest there are fruits as the source of food for the animals that live there. The forest can also maintain the local temperature...if we cut down all the trees the temperature will rise...so this helps to maintain the local temperature* | Generally, the students' knowledge of abiotic factors had improved. They demonstrated greater awareness of the abiotic factors in our environment and their functions |
| Threats | Students' knowledge of threats was limited. The reasons and explanations they gave were related only to human activity<br><br>Example of response—Respondent 1:<br>*Fish will die* | Students' ideas and knowledge relating to threats had increased. They were better able to explain threats to the environment related to habitats and humans<br><br>Example of response—Respondent 4:<br>*Animals that live in this habitat will die or they will lose their habitat. Animals that live there will not have water sources or food sources like fish in the river, and there will be no water way for transportation. Then there will be no income source for the fishermen and...* | Students' knowledge of threats had increased and they were able to give more explanation about threats to the environment |

**Acknowledgment** The authors wish to acknowledge the financial support received from the Exploratory Research Grant Scheme 2012 of the Ministry of Education Malaysia.



## References

Balgopal MM, Wallace AM (2009) Decisions and dilemmas: using writing to learn activities to increase ecological literacy. J Environ Educ 40(3):13–26
Epstein I, Stevens B, McKeever P, Baruchel S (2006) Photo elicitation interview (PEI): using photos to elicit children's perspectives. Int J Qual Met 5(3):1–11
Harper D (2002) Talking about pictures: a case for photo elicitation. Vis Stud 17(1):13–26
Hurworth R (2003) Photo-interviewing for research. Social Research Update 40. Accessed from http://sru.soc.surrey.ac.uk/SRU40.pdf
McRae L, Loh J, Bubb PJ, Baillie JEM, Kapos V, Collen B (2008) The living planet index – guidance for national and regional use. UNEP-WCMC, Cambridge
World Wildlife Fund (2002) Living planet report 2002
World Wildlife Fund (2008a) Building a sustainable future. WWF International, Gland
World Wildlife Fund (2008b) A roadmap for a living planet. WWF International, Gland, Switzerland
World Wildlife Fund (2012) Living planet report 2012. Biodiversity, biocapacity and better choices. WWF International, Gland

# Chapter 22
# Development of Interactive Multimedia Education System (IMES) as an International Education Platform

**Hiroshi Arisawa and Takako Sato**

**Abstract** In this paper we present an Interactive Multimedia Education System (IMES) that emphasizes the interactivity of distance education, i.e. it enables both instructors and students to see each others' faces, and enables high definition sharing of discussion materials with bi-directional laser pointer tracks enabling instructors and students to show what they are pointing at as they speak. A key characteristic of this system is comprehensive collaboration and equal participation from all sites connected to the course.

This paper explains the key technological elements of IMES in conducting multipoint, two-way remote courses with universities in Asia and Africa, and then describes out of these the most important "basic video transmission technology," "presentation material sharing system," "screen/pointer action sharing system," and "offline features including recording and editing."

**Keywords** Distance learning • E-learning • Interactive education • Multi-point communication

## 22.1 Introduction

In today's globalized society, research studies, surveys, and discussions are occurring across regions everywhere. In the environmental field especially, because it deals with phenomena of a global scale as well as region-specific issues, groups are commonly created for educational research that go beyond regional or national borders, as well as joint instruction, joint research, joint surveys, and research discussions.

H. Arisawa (✉) • T. Sato (✉)
Graduate School of Environment and Information Sciences, Yokohama National University, 79-7, Tokiwadai, Hodogaya-ku, Yokohama 240-8501, Japan
e-mail: arisawa@ynu.ac.jp; sugar@ynu.ac.jp

N. Kaneko et al. (eds.), *Sustainable Living with Environmental Risks*,
DOI 10.1007/978-4-431-54804-1_22, 

However, researchers that travel for courses or discussions are naturally faced with physical and financial limitations that force a reduction of frequency of these types of communication. For example, it is simply not possible to hold lectures with all participants at the same time every week, or have the type of mini-discussions or meetings that are so frequent in normal research laboratories, when spread across different countries and regions.

Meanwhile, the availability of Internet around the world and ICT technologies have helped to break down such location-specific barriers. One example of this is massive open online courses (MOOCs) (Broun 2013; Mackness et al. 2010) that make courses publicly available to many regions and persons; however, MOOCs do not have two-way capability, and only enable one-way transmission of courses for consumption. Also, MOOCs are best suited to courses with large numbers of students situated around the world, but are not so well suited for more average-sized courses such as those involving close interactions between students and instructors, such as courses that use group work, discussions or presentations to encourage a vigorous real time debate.

While many so-called e-learning tools have been proposed over the last several years (Keremidchieva and Yankov 2001; Nagataki et al. 2011; Nakabayashi 2011), we present now an Interactive Multimedia Education System (IMES) tool that emphasizes the interactivity described above, that enables instructors and students to see each others' faces, and enables high definition sharing of discussion materials, enabling lecture and discussion, while enabling each party to point at presentation materials (Realmedia 2013). This system enables two-way sharing of camera images and audio in real-time for both instructors and students. It also enables sharing presentation materials (such as PowerPoint presentation files) as well as laser pointer tracks, enabling instructors to even show students what they are pointing at as they speak. Students can also use laser pointers, enabling two-way communication for questions or discussion. A key characteristic of this system is enabling comprehensive collaboration and equal participation from all sites connected to the course.

We actually used IMES once a week to conduct remote lectures connecting nine universities in nine countries in Africa and Asia, clearly showing the system's effectiveness through actual multipoint, multi-region international lecture courses.

This paper explains the key technological elements of IMES in conducting multipoint, two-way remote courses with universities in Asia and Africa, and then describes out of these the most important "basic video transmission technology," "presentation material sharing system," "screen/pointer action sharing system," and "offline features including recording and editing."

## 22.2 Key Technical Elements of IMES

Figure 22.1 shows the basic transmit/receive concept of IMES.

Two screens are essentially provided for each site, one screen for displaying the camera images of the instructor and other course sites at the same time, and another screen for displaying PowerPoint or other presentation materials used for the course.

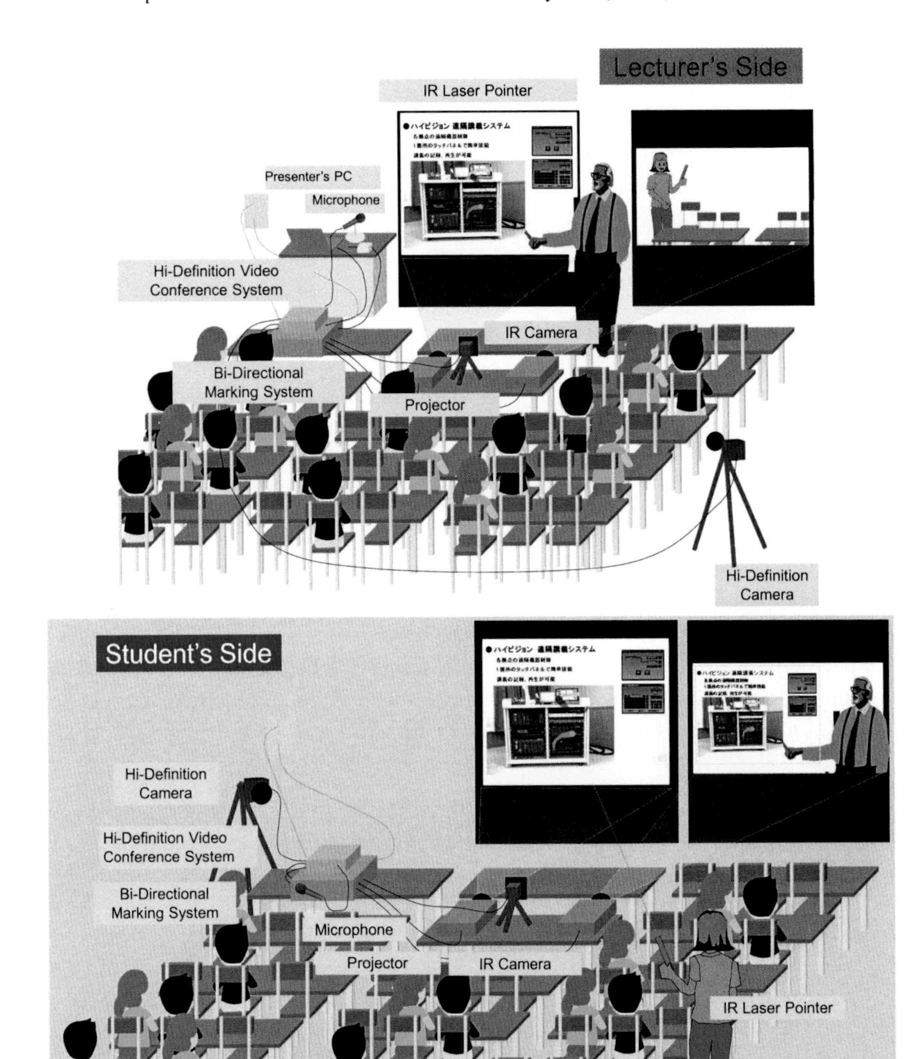


**Fig. 22.1** The basic transmit/receive concept of IMES

IMES is comprised of the following four basic technologies. Items in brackets [] are for multipoint use.

1. Enable two-way camera video between instructors and student(s) [video from all sites is combined and shared on a single screen]

2. Students can hear instructors, but instructors can also hear audio of student questions [if multiple parties are speaking at once, audio from all sites is heard as mixed together]
3. Presentation materials (PowerPoint, etc.) can be shared with sufficiently high screen quality that even small print or graphics are legible
4. It is also possible to share where a student or instructor is pointing at on presentation materials [all sites can use pointers at the same time]

High definition video conferencing system products have come onto the market in recent years (Sony 2008), and have enjoyed good adoption rates for uses such as connecting branch offices of companies. However, they are not well suited in some ways for use in courses connecting multiple locations for educational institutions such as universities. For example, the following can be cited as problems in realizing the four key elements mentioned above.

- The system must be work over normal Internet connections that are available to anyone, but the reality is that some Internet connections are fast, some are slow, and some are very slow. The system must work over slow Internet connections between different universities in different countries.

  (Commercial videoconferencing systems presume the availability of higher speed circuits than are generally available at universities, and there is little consideration for sharing presentations such as PowerPoint. For making a presentation, for example, even though a high speed connection is required, the graphics quality is typically poor, and gets worse for multipoint conferences.)
- Specialized equipment is generally required for connecting more than ten locations (MCU—Multiple Communication Unit), which is extremely expensive.

In other words, a system is needed that overcomes these issues while realizing points (1) to (4) above.

## 22.3 Basic Video Transmission

Video is one of the most basic and persuasive tools for sharing information, but because it consumes so much bandwidth, there are technological difficulties to its successful implementation, with much research and development on how to overcome this.

Because large volumes of data (video/audio) travel in one direction, or in multiple directions to multiple destinations over the Internet, the basic technological requirement is to be able to compress, transmit and play back in real time.

Conventional analog television transmission resolution is equivalent to 640×480 pixels when shown on a computer screen. This is normally transmitted at an actual rate of 30 frames/s. High definition television transmits at 1,920×1,080 pixels (varies slightly depending on codec used) at 30 frames/s this—a dramatic increase in video clarity.

But the data size of such video is extremely large. It requires three bytes to represent each one of these pixels in natural color at 16.77 million colors. This results in a basic calculation of 216 Mbits (megabits)/s to transmit just 1 s of conventional television, or 1,458 Mbit/s for high-definition television (HDTV)—much higher than the transmission rate of a typical Internet connection (100 Mbit/s for even the fastest circuits). For this reason, technologies are needed that compress the original image with an acceptable loss of image information. Digital terrestrial broadcasts can be recorded in high definition today using the HDV format at a rate of 25 Mbits/s, and this can be further reduced to 2–8 Mbits/s using the H.264 profile included in recent-model hard disk video recorders, enabling encoding/decoding video that appears natural to the human eye. The Internet connections that are typically available to us in overseas course locations are in the range of 1–3 Mbits/s transmission, so it is this type of advanced encoding/decoding technology that has made it possible to send HDTV-grade video in real time over the Internet.

We had originally developed a system that used proprietary encoding/decoding, but with the emergence on the market of affordable, good performance video conferencing systems based on H.264, the current IMES promotes the use of commercially available products.

However, with up to about 12 parties simultaneously connecting around the world in international courses, the problem arose of not reaching the point of purchasing the high-cost MCU previously mentioned, but not being able to meet the requirements with a video conferencing system that includes multipoint communication functionality (up to about six locations). We therefore developed a method for multi-site communication for up to 20 locations, by connecting relatively affordable equipment able to relay up to six locations, and then using a specialized PC to control these externally.

## 22.4 Presentation Material Sharing System

Presentation materials such as PowerPoint slides are commonplace in university course environments. Especially in science-related courses, there may frequently be descriptions written in small characters, formulae with suffices, or detailed drawings or photographs. Therefore, when conducting courses across multiple locations, it is necessary to continually deliver and share high quality presentation images from the lecture site to all student sites. However, sending presentation images containing large amounts of data places puts a heavy burden on networks. For example, the XGA resolution typically used when showing a PowerPoint presentation on a screen with a projector is 1,024 by 768 pixels. If the color information for each pixel requires three bytes, then the entire screen requires 19 Mbits. While this is less than for video, it is impossible to continually send presentations, smoothly turn pages and play animations over the Internet (an animation requires at least 10 frames/s). The even higher SXGA resolution is growing in popularity—this resolution requires 32 Mbits for one screen. Compression technologies can be used, but visual artifacts

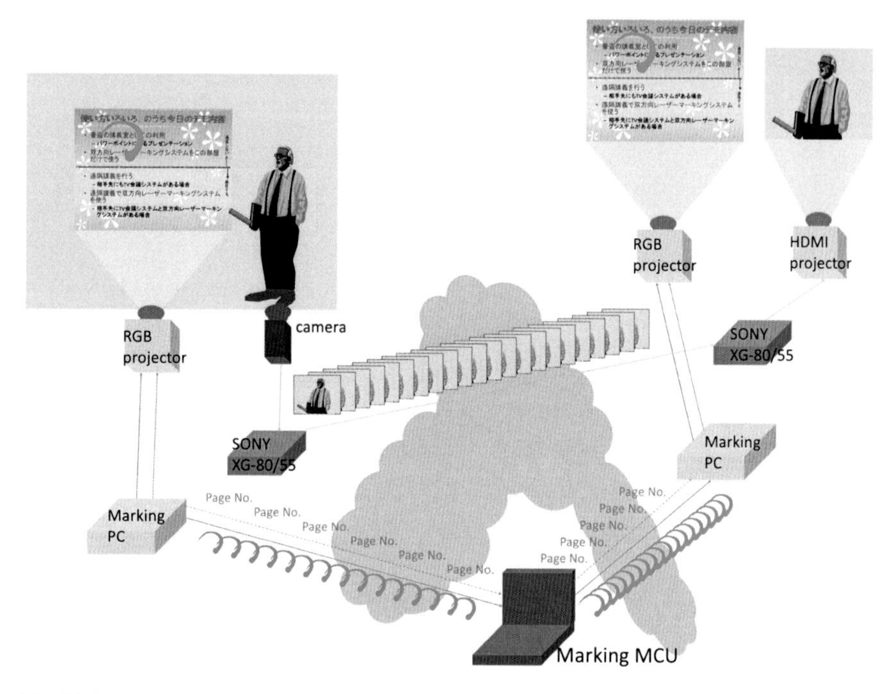


**Fig. 22.2** The conceptual diagram of Mode-B presentation

are more obvious when compressing presentations than videos, preventing the use of high compression rates.

In order to resolve this issue, the method that we developed for IMES sends shared presentation materials just before the course. The only data sent in real time during the course is page turns and laser pointer traces, which is a lot less data. However, the necessary presentation images for the entire course must be effectively delivered to all course locations that are joining that course. We solved this problem by setting up a simple server called a Presentation Multipoint Control Unit (PMCU). Course presentation materials are uploaded as files to this server in advance, and each student site can download the server files at any time before the course. The current PMCU is a simple software that can run on a laptop computer, and is of sufficiently high performance to support more than 20 delivery sites.

We call this presentation sharing method "Mode-B" presentation.

A conceptual diagram is shown in Fig. 22.2.

Presentations are converted to data files and pre-sent using Mode-B either using an image capture device from the instructor's PC to import to the control PC, or the original source data being converted to data file directly if it has PowerPoint or other specific file format. The former method supports any presentation software and can even convert animations into frame sequences, enabling simple sharing of animated presentations to all sites.

Mode-B not only prevents degraded screen quality of presentation materials, but also frees network bandwidth for camera video and audio because there is no load on the network for sending presentation images during remote learning, improving overall quality.

Mode-B is an extremely efficient and reliable communication method, but sometimes during specialized courses or discussions, it becomes necessary to also transmit text written on a whiteboard or images presented on a physical projector. Instructors sometimes also need to necessary to urgently share small presentations created on the spot for announcements or the like. To address these requirements, IMES also provides "Mode-C" to directly acquire images of cameras, physical projectors or instructor computers and "slowly" distribute them to each connected site as to not put a heavy load on the network.

## 22.5 Screen/Pointer Action Sharing System

Remote learning or specialized field discussions frequently use electronic media or presentation software (PowerPoint, etc.), requiring interaction through actions such as directly pointing at or writing on part of the screen. Video conferencing systems are designed mainly for environments where (printed) presentation materials can be shared in advance, and do not tend to emphasize interaction through the screen. However, most instructors are only able to prepare materials for that day or the previous day at most, and interaction is extremely important for classes or courses where they point at the screen that day, sometimes asking questions or involving students in a discussion (university professors especially prefer to use laser pointers to explain things while looking at students).

It is desirable to share a single screen across the network to multiple sites, allow any site to mark the screen using laser pointers, and enable all sites to see this at the same time. However, it is surprisingly difficult to share presentation screens over the network using personal computers. While it would be simple to just point a camera at the PC screen and transmit that video, the image becomes blurred and distorted when compressed and decompressed, and is not suitable for showing a PC screen in fine and clear detail. For this reason, modern video conferencing systems import and send video signals directly from PCs (the H.264 compression standard is tuned for presentations in this case, or a specialized H.239 compression/transmission technology is used). However, it is difficult to share marks in such a case. In other words, when the instructor is using a laser pointer or pointing stick, sites other than the instructor site cannot see where the instructor is pointing on the screen. Other methods exist such as moving a mouse cursor on a screen or making marks using a tablet device, but these limit the ability of instructors to move about and are generally not preferred.

To address this problem, a "multipoint laser marking system" was developed for IMES. The concept is shown in Fig. 22.3.

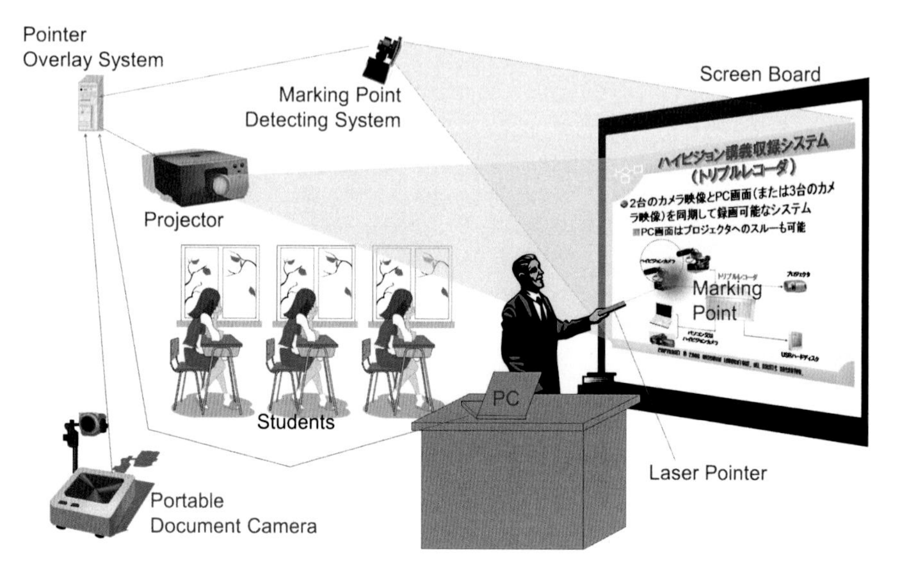


**Fig. 22.3** The concept of multipoint laser marking system

The actual process is as follows. The most significant point is that an infrared (IR) laser pointer is used, whose light is not visible to the human eye.

1. The instructor points use a special IR laser pointer to point at the screen.
2. An IR camera detects the laser spot location on the screen.
3. A control PC reads the laser spot and creates a marking overlay screen (a transparent screen on which the pointer track or other marks are drawn), based on the IR spot coordinates.
4. The control PC imports the instructor's PC screen and overlays the transparent marking screen from (3) above.
5. For Mode-B, the currently displayed presentation screen's ID number and marking information (pointer location, color, track length) are transmitted to each student site. For Mode-C, the presentation screen and marking information are transmitted.
6. By the same process as (4) above, pictures are created at student sites from overlaid presentation images and marking images, and marking screens from each other site are also overlaid and displayed on all site screens.

This enables an environment in which the presentation screen of one site is presented at the same time to multiple student sites, and any site can also mark on the screen. Figure 22.4 shows a screen example for one on one two-way marking. This system enables instructors and students, even when remote, to engage in discussion and share markings, as if they are looking at a single screen. We call this Shared Screen, and are actively promoting in remote courses. It is regarded as being extremely effective by instructors and students alike.

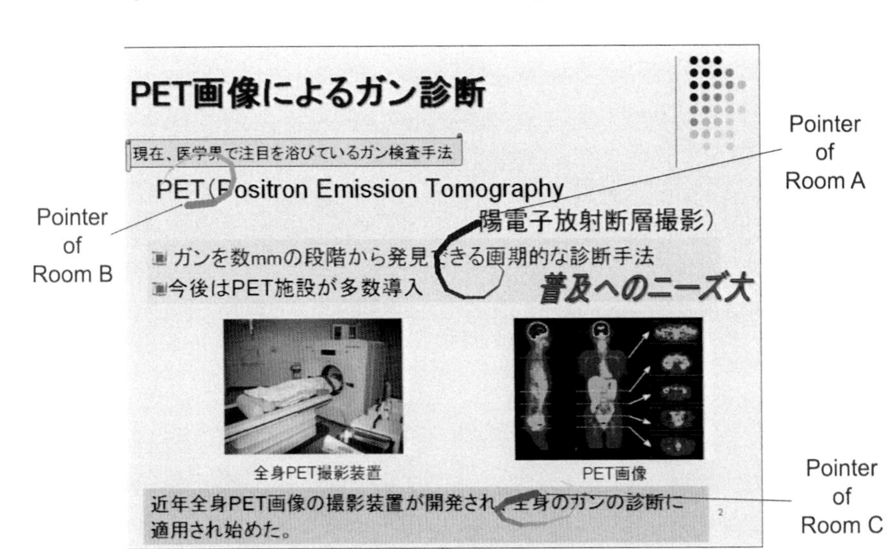


**Fig. 22.4** A screen example for one on one two-way marking

## 22.6 Offline Features Including Recording and Editing

As described above, IMES is especially effective when one course or class is shared among multiple locations. However, there are cases where students or instructors want to record lectures and distribute them later (the international courses we observed included participants from nine countries in locations such as Asia and Africa, with as many as 6 h time zone difference. They somehow made do by starting class at 4pm. This will become impossible if any participants join from North or South America).

Therefore, we also developed as an external device for IMES a system to enable recording and storing lectures in high resolution, and making edits, as needed, for standalone transmission at a later time. The problem here is that IMES sends and receives two screens—video of the instructor/students and the presentation screen. When recorded as two separate video streams, both timelines must be kept in careful sync, or editing becomes extremely difficult when importing into editing software (also called non-linear editing). There is also no video or audio for the presentation screen, because it is difficult to shift the time of one in the editing software and align it with the other. To address these problems, we developed a simultaneous video recording tool.

This simultaneous video recording tool saves the two video types as separate files on to a single hard disk, and keeps their timelines in sync when they are imported into editing software as content for editing. During the editing process, they are typically combined into a single widescreen image (the 16:9 aspect ratio used in high definition broadcasts). During presentation, the screen on the right is

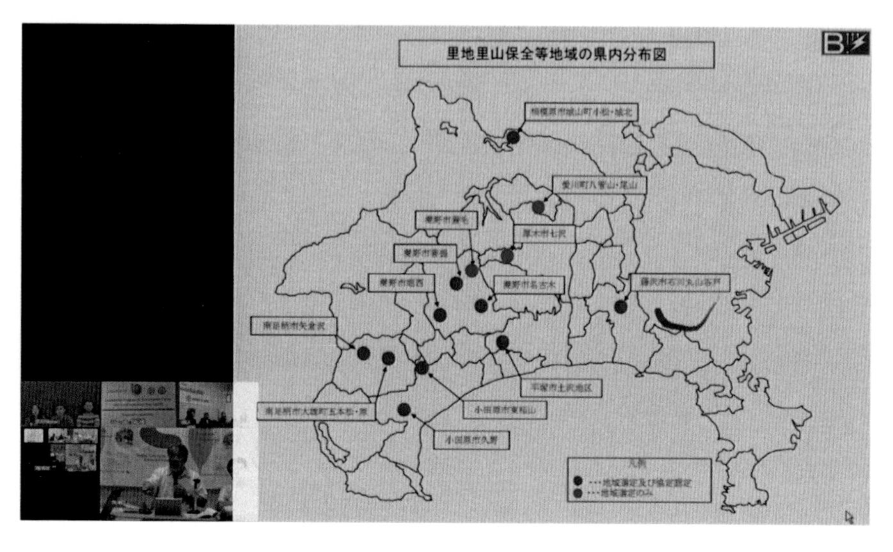


**Fig. 22.5** An example of an edited screen

the original presentation screen (4:3 aspect ratio), and laser pointer markings are embedded as images. Instructor video other video is placed in the space left over at the left side of the screen. Of course, other editing is possible as well, such as swapping the two screens depending on the course/class, or using wipe or other transition effects. An example of an edited screen is shown in Fig. 22.5 above.

Users love this combined screen view, so as an added feature, IMES currently makes available a "low-bitrate A/V distribution system" as a web view (at a lower resolution) at all times. It is convenient when the signal is lost during a class or to monitor class conditions.

## 22.7 Conclusion and Topics for Future Discussion

As described above, we developed and continually improved a system under the concept of multi-point, real time and bidirectional courses, actually using the system throughout the year in a course held once a week, by placing equipment in eight overseas countries. The IMES system we built was comprised of commercially available video conferencing systems and a control PC with the remote presentation system software described above, using one of three configurations from rack mount to laptop PC, depending on the time of configuration and size of the site (Fig. 22.6).

IMES was proven to be extremely effective through a series of international course experiences. Field work is extremely important in environmental studies as was the case here, and the system showed its value in the importance of discussing things before field work, such as sharing opinions or going over detailed schedules

**Fig. 22.6** (**a**) Rack mount type; (**b**) Desktop PC type; (**c**) laptop PC type of External appearance of IMES

**Fig. 22.7** Several class scenes using IMES

or procedures, and for summarizing or presenting post-field work data. The field work of an international team and pre- and post-field work discussion work went very smoothly, thanks to the use of IMES.

IMES is currently being evaluated in classes other than environmental studies as described above, such as at the medical faculty of a nearby university (not our own university), being used to hold seminars with remote graduate students in one professor's laboratory, as well as being used for non-recurring events such as symposiums.

Several class scenes are shown in Fig. 22.7 above.

However, the skillful use of IMES depends not just on student not simply feeling he or she is watching a "video course," but how skillfully the instructor can bring the people from the remote side closer to his or her side. Instructors should call on many students and invite them to speak even when they don't volunteer, announce the results of group work, showing the importance of the strength of course coordinators in making the course more interactive.

Only pointer tracks can be shared at this time, but there remain many development desires, such as sharing over an electronic board instead of whiteboards, simultaneous control of various software packages in every location, or enabling presentation of even more types of presentation materials.

In any case, it is without a doubt that communication tools such as IMES that break down space and time barriers will grow even more essential the global society of tomorrow.



## References

Broun JT (ed) (2013) Research & practice in assessment – special issue: MOOCs & Technology 8. Virginia Assessment Group, Virginia

Keremidchieva G, Yankov P (2001) Challenges and advantages of distance learning systems. Inform Secur 6:115–121

Mackness J, Mak SFJ, Williams R (2010) The ideals and reality of participating in a MOOC. In: Proceedings of the 7th international conference on networked learning 2010, Hvide Hus Hotel, Aalborg, Denmark

Nagataki H, Noguchi K, Katsuma R, Yamauchi Y, Shibata N, Yasumoto K, Ito M (2011) A distance learning system with customizable screen layouts for multiple learning situations. In: 3rd International conference on computer supported education (CSEDU 2011), Noordwijkerhout, The Netherland, pp 94–102

Nakabayashi K (2011) e-Learning technology standard promotion activities in Japan. In: Proceedings of 24th ISO/IEC JTC1 SC36 Plenary and WG meetings, International Open Forum, pp 63–66

Realmedia Lab (2013) Manual of interactive multimedia education system (IMES). Realmedia Lab Corporation, Yokohama

Sony (2008) HD visual communication system IPELA PCS-XG80/XG80S operating instructions (version 2.0). Sony Corporation, Tokyo

# Index




N. Kaneko et al. (eds.), *Sustainable Living with Environmental Risks*,
DOI 10.1007/978-4-431-54804-1, © The Author(s) 2014